D0828441

WAVELET THEORY AND ITS APPLICATIONS

THE KLUWER INTERNATIONAL SERIES IN ENGINEERING AND COMPUTER SCIENCE

VLSI, COMPUTER ARCHITECTURE AND DIGITAL SIGNAL PROCESSING

Consulting Editor
Jonathan Allen

Latest Titles

Neural Models and Algorithms for Digital Testing, S. T. Chakradhar, V. D. Agrawal, M. L. Bushnell,
ISBN: 0-7923-9165-9

Monte Carlo Device Simulation: Full Band and Beyond, Karl Hess, editor
ISBN: 0-7923-9172-1

The Design of Communicating Systems: A System Engineering Approach, C. J. Koomen
ISBN: 0-7923-9203-5

Parallel Algorithms and Architectures for DSP Applications, M. A. Bayoumi, editor
ISBN: 0-7923-9209-4

Digital Speech Processing: Speech Coding, Synthesis and Recognition A. Nejat Ince, editor
ISBN: 0-7923-9220-5

Sequential Logic Synthesis, P. Ashar, S. Devadas, A. R. Newton
ISBN: 0-7923-9187-X

Sequential Logic Testing and Verification, A. Ghosh, S. Devadas, A. R. Newton
ISBN: 0-7923-9188-8

Introduction to the Design of Transconductor-Capacitor Filters, J. E. Kardontchik
ISBN: 0-7923-9195-0

The Synthesis Approach to Digital System Design, P. Michel, U. Lauther, P. Duzy
ISBN: 0-7923-9199-3

Fault Covering Problems in Reconfigurable VLSI Systems, R.Libeskind-Hadas, N. Hassan, J. Cong, P. McKinley, C. L. Liu
ISBN: 0-7923-9231-0

High Level Synthesis of ASICs Under Timing and Synchronization Constraints D.C. Ku, G. De Micheli
ISBN: 0-7923-9244-2

The SECD Microprocessor, A Verification Case Study, B.T. Graham
ISBN: 0-7923-9245-0

Field-Programmable Gate Arrays, S.D. Brown, R. J. Francis, J. Rose, Z.G. Vranesic
ISBN: 0-7923-9248-5

Anatomy of A Silicon Compiler, R.W. Brodersen
ISBN: 0-7923-9249-3

Electronic CAD Frameworks, T.J. Barnes, D. Harrison, A.R. Newton, R.L. Spickelmier
ISBN: 0-7923-9252-3

VHDL for Simulation, Synthesis and Formal Proofs of Hardware, J. Mermet
ISBN: 0-7923-9253-1

WAVELET THEORY AND ITS APPLICATIONS

by

Randy K. Young
Pennsylvania State University

KLUWER ACADEMIC PUBLISHERS
Boston/Dordrecht/London

Distributors for North America:
Kluwer Academic Publishers
101 Philip Drive
Assinippi Park
Norwell, Massachusetts 02061 USA

Distributors for all other countries:
Kluwer Academic Publishers Group
Distribution Centre
Post Office Box 322
3300 AH Dordrecht, THE NETHERLANDS

Library of Congress Cataloging-in-Publication Data

Young, Randy K., 1963-
Wavelet theory and its applications / Randy K. Young
p. cm. -- (Kluwer international series in engineering and computer science ; SECS 189)
Includes bibliographical references and index.
ISBN 0-7923-9271-X
1. Signal processing--Mathematics. 2. Wavelets (Mathematics)
3. Linear time invariant systems. I. Title. II. Series
TK5102 . 5 . Y67 1992
621 . 382 ' 2--dc20 92-23362
CIP

Printed on acid-free paper.

Printed in the United States of America

To: *Rita, Jason, and Dad*

Contents

Foreword

The continuous wavelet transform has deep mathematical roots in the work of Alberto P. Calderon. His seminal paper on complex method of interpolation and intermediate spaces provided the main tool for describing function spaces and their approximation properties. The Calderon identities allow one to give integral representations of many natural operators by using simple pieces of such operators, which are more suited for analysis. These pieces, which are essentially spectral projections, can be chosen in clever ways and have proved to be of tremendous utility in various problems of numerical analysis, multidimensional signal processing, video data compression, and reconstruction of high resolution images and high quality speech. A proliferation of research papers and a couple of books, written in English (there is an earlier book written in French), have emerged on the subject. These books, so far, are written by specialists for specialists, with a heavy mathematical flavor, which is characteristic of the Calderon-Zygmund theory and related research of Duffin-Schaeffer, Daubechies, Grossman, Meyer, Morlet, Chui, and others.

Randy Young's monograph is geared more towards practitioners and even non-specialists, who want and, probably, should be cognizant of the exciting proven as well as potential benefits which have either already emerged or are likely to emerge from wavelet theory. For example, the theory of multidimensional processing of video signals is a subject of very current interest and a Special Issue has, recently (in May 1992), been devoted to the topic in the journal on *Multidimensional Systems and Signal Processing*, also published by Kluwer Academic Publishers. The multiresolution method decomposes high resolution images into a hierarchy of pieces or components, each more detailed than the next. Thus, the multiresolution approach, which is based on the wavelet theory of successive approximations provides the mechanism for transmitting various grades of images depending upon the quality of reconstruction sought and the limitation set by channel capacity as well as the need for compatibility with existing receivers when high definition television, for example, enters the market. Certainly, the single resolution approach is incapable of adjusting to the practical constraints described. The myriad of applications which wavelet theory has spawned will, undoubtedly, stimulate the writing of many books which will emphasize applications and attempt to widen the circle of readers. I am glad that Randy Young has taken the initiative to do just that, and I believe that others will follow his footsteps.

N. K. Bose
The Pennsylvania State University
University Park

Preface

This book extends wavelet theory, both theoretically and practically. But even the theoretical extensions are directed toward practical applications. The theory and derivations require only college senior level mathematics. Primarily, signal processing applications are addressed, but these applications are quite general (i.e., signal and system models) and can be employed in many diverse fields.

Initially, in Chapter 1, wavelet theory is introduced and terminology is defined. The time scaling action of the wavelet transform is emphasized. A pictorial demonstration of the time scaling action is presented by creating several scaled and translated "mother wavelets." After some brief motivational comments, the structure of the wavelet transform is provided along with a pictorial example.

Chapter 2 considers both continuous and discrete wavelet transforms and reviews established wavelet theory. The review is not detailed but references are cited if additional details are desired. Since this book concentrates on nonorthogonal and "sophisticated" wavelets, the special cases of orthogonal, biorthogonal, and multiresolution wavelet transforms are presented, but with an emphasis on their constraints and limitations. These special wavelet transforms are extremely efficient and very powerful (and these special wavelet transforms are considered "wavelet theory" for many people). This book addresses the wavelet theory that is not included within these special wavelet transforms and relaxes the constraints introduced by these special transforms. Very general mother wavelets will be required for the wideband or wavelet system models (the input signal will be employed as the mother wavelet).

After justifying these more general wavelet transforms, the properties of these transforms are investigated. Chapter 3 primarily examines the gain and resolution of the wavelet transforms by relating the processing to matched filtering concepts. The wideband or wavelet processing is also compared to the Fourier or narrowband processing. Throughout this book, the parameterization of the wavelet transform by the mother wavelet is of critical concern. The properties of these mother wavelets are most easily described by ambiguity functions and, thus, ambiguity functions are described first. After discussing ambiguity functions, wideband ambiguity functions are efficiently formulated with wavelet transforms. This reformulation produces a cross wavelet transform that extracts the common features of the two signals (or systems, channels, etc.) being cross processed.

Then, in Chapter 4, the "sophisticated" mother wavelets are compared to their tone-like counterparts from an ambiguity function point of view. By considering multiple mother wavelets, efficient wavelet domain representations can be formed. The computation of multiple wavelet transforms is efficiently reformulated by a new operator, the mother mapper operator. This operator efficiently maps a wavelet transform with respect to one mother wavelet to a new wavelet transform with respect to a different mother wavelet.

Next, in Chapter 5, an original model for space-time-varying (STV) systems, including linear time-invariant systems as a special case, is formulated with wavelet theory. The wavelet transform utilized for this system operator is consistent with the wavelet transform definition employed throughout the world. This system model allows the system to be characterized by an "energy" distribution. A new operator, the STV wavelet operator, is defined and its properties are examined. Besides modelling more general systems, this operator allows both wideband and nonstationary input and output signals. However, the new system representation is limited by acceleration(s) in the system (analogous to the narrowband system model being limited by velocity(s)). By employing the STV model, the system model can be estimated over a longer time period than conventional system estimators (thus providing more gain and better resolution). The longer estimation interval is achieved by accounting for the system's time variation over the estimation interval.

The new system model can easily represent a reflection from a fast moving, non-accelerating object (the wavelet system model can approximate the first order time mapping of the reflection process); the new system can represent true time scaling. The representation for a single fast moving object is a single point or impulse (or delta function) in a scale-translation plane. Since a fast moving object can be properly modelled by just an impulse in the plane, extremely general systems can be modelled by exploiting all of the two-dimensional freedom that the new system representation offers (LTI systems only allow the one-dimensional impulse response to represent a system). Time-frequency models cannot easily model this reflection process or time scaling operation.

The properties of the new operator are stated and justified. By using these properties, the STV wavelet operator naturally maps a wavelet transform of the input signal to a wavelet transform of the output signal; however, the transform domain mapping is not just a multiplication (as it is for the Fourier system frequency response). A comprehensive example demonstrates the new operator's application. Due to the generality of this system model, it can be employed in a wide variety of fields. The different requirements for modelling systems instead of signals is detailed with an emphasis on resolution requirements.

Finally, in Chapter 6, the practical application of imaging or scattering characterization is addressed and modeled as a system identification problem. Several forms of the STV wavelet operator are utilized to provide a general, yet efficient, solution. Practical considerations in implementing this model are examined and comparisons to analogous narrowband and stationary models are made.

WAVELET THEORY AND ITS APPLICATIONS

Chapter 1: Introduction/Background

Wavelet theory is the mathematics associated with building a model for a signal, system, or process with a set of "special signals." The special signals are just little waves or "*wavelets*." They must be oscillatory (waves) and have amplitudes which quickly decay to zero in both the positive and negative directions (little). See Figure 1.1 for an example of a wavelet (this is a classical wavelet, termed the "Morlet mother wavelet," after its inventor). The required oscillatory condition leads to sinusoids as the building blocks (see Figure 1.2). The quick decay condition is a tapering or windowing operation (see Figure 1.3). These two conditions must be simultaneously satisfied for the function to be a little wave or wavelet. Forming the product of the oscillatory and decay functions yields the wavelet of Figure 1.1.

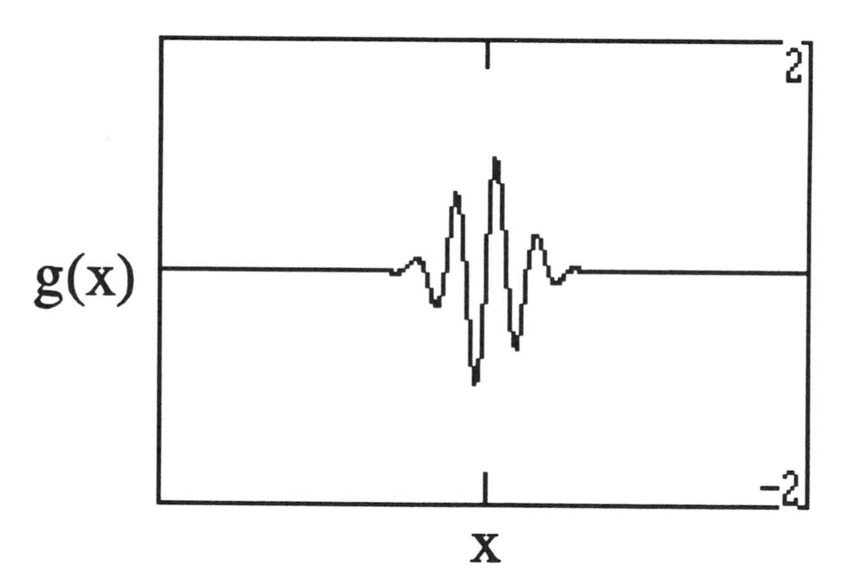

Figure 1.1: Morlet Mother Wavelet

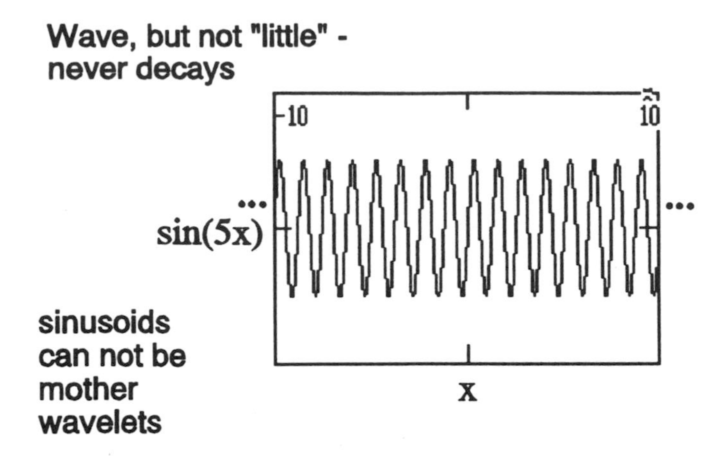

Figure 1.2: Oscillatory or Wave Requirement

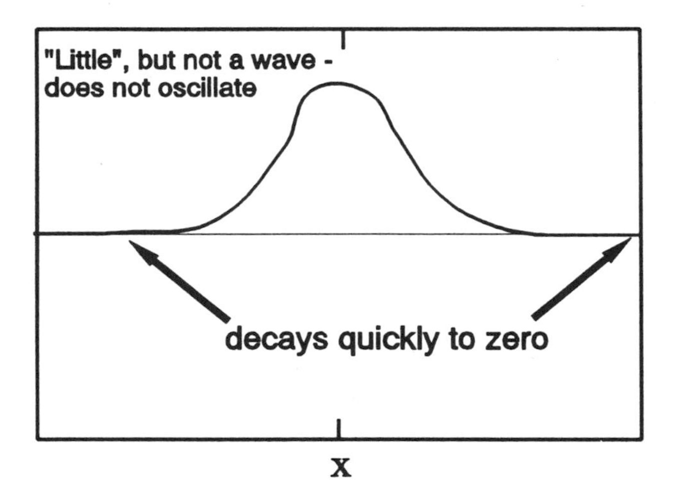

Figure 1.3: "Little" Requirement

Sets of "wavelets" are employed to approximate a signal (or process, or system, etc.) and each element in the wavelet set is constructed from the same function, the original wavelet, appropriately called the *mother wavelet*. Each element of the wavelet set is a scaled (dilated or compressed) and translated (shifted) mother wavelet. Figure 1.4 and Figure 1.5 display several elements of such a set corresponding to the Morlet mother wavelet in Figure 1.1. Note the constant shape of these scaled and translated functions; the same number of oscillations is in each wavelet or wave packet. Also, note that the scaled wavelets include an energy normalization term, $\frac{1}{\sqrt{a}}$, that keeps the energy of the scaled wavelets the same as the energy in the original mother wavelet. When "scaling" is used in this book it will implicitly include this normalization term (the mother wavelet is strictly being scaled and multiplied by a constant or amplified but the term "scaling" will be used instead of constantly repeating "normalized scaling").

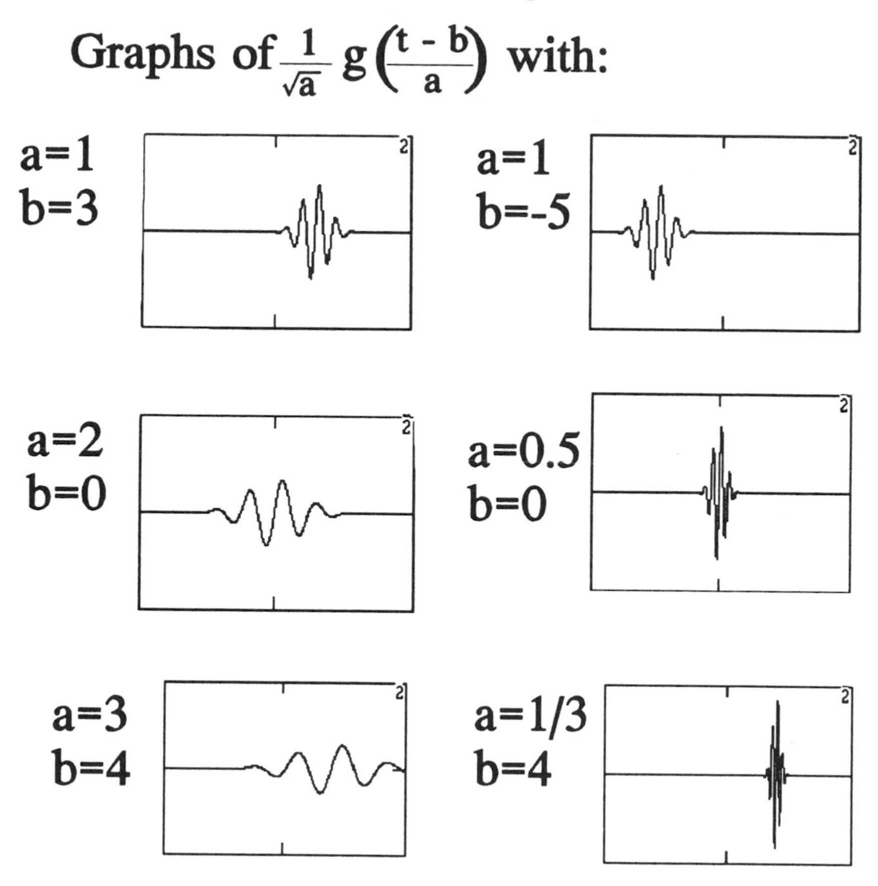

Figure 1.4: Scaled and Translated Mother Wavelets

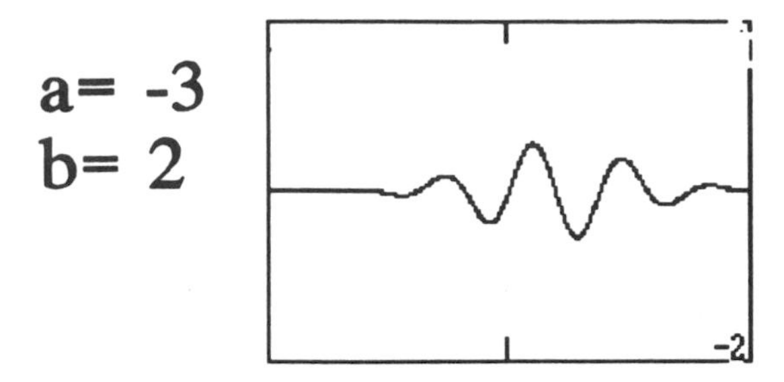

Figure 1.5: Negatively Scaled Mother Wavelet

Wavelet theory can be employed in many fields and applications, such as image analysis, communication systems, biomedical imaging, radar, air acoustics, theoretical mathematics, control systems, and endless other signal processing areas. However, *access to wavelet theory has been limited* because much of the wavelet literature requires that the reader understand the mathematics of functional analysis, group representation theory, Hilbert space theory, the Frobenious-Schur-Godement Theorem, etc. One purpose of this book is to present the wavelet theory so that it is *accessible to a broader audience than the current readers of the research* and, thus, the mathematical derivations in this book are concise and complete yet do not require knowledge of a broad range of mathematics (at the sacrifice of some mathematical rigor). Pictures are presented not just for simulation results but also for visualization of the concepts and operations. Besides trying to make wavelet transforms more applicable, the book also provides new results that extend and generalize wavelet theory.

Wavelet Theory Basics - Scaling and Translation

Just as numbers may represent objects, wavelet theory is another technique for representing "things." The "things" wavelet theory can represent include stock market prices, solutions to differential equations, a heart beat, the echo process, blood flow through the body, a meteorological model, and many other "things." Besides representing objects, wavelet theory can be exploited to represent actions or operations (systems) as well. Wavelet theory is a mathematical tool that can be applied almost anywhere and like most tools, its primary purpose is to improve the efficiency (analogous to a wrench turning a bolt rather than using your fingers). Although numbers represent objects, many representations for numbers exist (i.e., decimal, binary, etc.) and the representation that is chosen is dictated by its efficiency in a particular application (the decimal representation for accounting and the binary or hexadecimal representation for computer designs). Similarly, the wavelet theory representations are more efficient than alternative representations for some applications and, thus, wavelet theory should be employed there. In some applications wavelet theory may produce undesirable inefficiencies and, thus, wavelet theory should not be blindly forced upon an application.

Wavelet theory represents "things" by breaking them down into many interrelated component pieces, similar to the pieces of a jig-saw puzzle; when the pieces are scaled and translated wavelets, this breaking down process is termed a *wavelet decomposition or wavelet transform*. *Wavelet reconstructions or inverse wavelet transforms* involve putting the wavelet pieces back together to retrieve the original object or process. Wavelet theory consists of the study of these pieces (wavelets), their properties and interrelationships, and how to put them back together.

Wavelet theory involves the scaling or warping operation. For example, watching a movie on your TV with the VCR running on fast forward is the scaling (time scaling) action. A function (the movie) has been time scaled (fast forwarded) and a time-compressed function (movie) is created. Although the time-compressed movie is still the same movie, its representation in terms of spatial-temporal parameters has changed (assuming no degradation is acceptable, to send the time-compressed (shorter duration) movie over a satellite link would require several satellite channels instead of just one satellite channel for the uncompressed version). The information content of the movie does not change. The scaling action alters the independent variable (time in this example) to create new functions that have the same structure except they are either expanded or compressed (see Figure 1.4 and Figure 1.5).

Wavelet theory also involves the *translation* or shifting operation. Time delays are typical "translations" of the time axis or variable. For example if the movie is shown at 5 pm instead of its original 2 pm air-time, then it is shifted or translated by 3 hours.

If the *translation operation is combined with the scaling* action then the composite operation is referred to as an *affine operation*. The affine operation simultaneously scales and translates the independent variable. An example of the affine operation in the movie example is a fast forwarded version (scaled) of the movie shown at 5 pm (translated 3 hours from 2 pm). Mathematically, the affine operation maps the original independent variable, x, into a new independent variable, x', by the formula:

$$x' = \frac{x-b}{a} \tag{1.1}$$

where "a" is the scale and "b" is the translation. If your VCR ran twice as fast, then the scale would be $a = \frac{1}{2}$. The translation is not as straight forward. To correctly map the independent variables, a *time reference must be considered whenever a scaling operation is applied*. To avoid being "side tracked," the time referencing is further investigated in Appendix 1-A. The complication due to time referencing should be noted - time referencing is the requirement that separates time invariant systems from time-varying systems. Systems that involve time scaling will be time-varying systems and will require a time reference. In addition, the affine operation is not commutative; when the scaling and translation are applied in reverse order, a different operation results.

In wavelet theory the scaling and translation operators act simultaneously on the *mother wavelet* function. The name "wavelet," literally meaning "little wave," originated from the study of short duration, seismic acoustic wave packets [Gro1]. The mother wavelet (the initial little wave) is the kernel of the wavelet transform (detailed later). Performing affine operations on the mother wavelet creates a set of scaled and translated versions of this mother wavelet function. This set of scaled and translated mother wavelets is called a *wavelet set* (see Figure 1.4 and Figure 1.5).

Mathematically, as detailed in Chapter 2, the continuous wavelet transform (decomposition) of a function, f(x), with respect to a mother wavelet, g(x), is:

$$\mathbf{W}_g[f(x)](a,b) = |a|^{-\frac{1}{2}} \int f(x)\, g^*\left(\frac{x-b}{a}\right) dx = \mathbf{W}_g f(a,b) \quad (1.2)$$

where the superscript "*" denotes complex conjugation, "a" is the scale parameter, "b" is the translation parameter. The notation $\mathbf{W}_g f(a,b)$ will be used for brevity but it should be clear that the original function is over the independent variable, x, and is mapped to a new two-dimensional function across scale and translation, a and b, respectively. A wavelet coefficient, $\mathbf{W}_g f(a,b)$, at a particular scale and translation represents how well the signal, f, and the scaled and translated mother wavelet match; if the signal, f, is similar to the scaled and translated mother wavelet, then the coefficient will have a big magnitude. The coefficient represents the "degree of correlation" between the two functions at a particular scale and translation. The set of all wavelet coefficients, $\mathbf{W}_g f(a,b)$, are the *wavelet domain representation* of the function f *with respect to the mother wavelet g*. The affine operations represented by the scale and translation parameters, (a,b), act on the argument, x, of the mother wavelet (for a specific mother wavelet and several specific scales and translations, the warped mother wavelets were shown in Figure 1.4 and Figure 1.5). A representative picture of a signal and its wavelet transform is shown in Figure 1.6. Note that the one-dimensional signal is now represented by a two-dimensional surface. (Although Figure 1.6 only shows a "real" wavelet domain representation, for general complex representations both the magnitude and phase of the transform would be required.) The mathematical details and rigorous justification of this transform are subsequently provided in Chapter 2.

Wavelet Transform Mapping

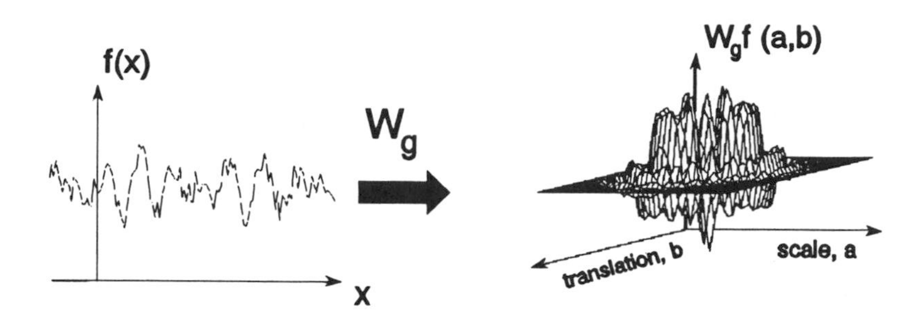

Figure 1.6: Wavelet Transform of f with respect to g

The wavelet transform can be related to the more commonly used Fourier transform or Fourier series. The Fourier models represent functions as weighted sum of exponentials at different frequencies. The weight at each different frequency is the Fourier coefficients. Wavelet models analogously represent functions as a weighted sum of scaled and translated mother wavelets. The wavelet transform has a mother wavelet replace the exponential, scaling and translation replace frequency shifting, and a two-dimensional surface of wavelet coefficients replace the one dimensional Fourier coefficients. As a special case where the mother wavelet is $g(x) = e^{jx}$, $a = \frac{1}{\omega}$, and $b = 0$, then the wavelet transform in equation (1.2) becomes:

$$W_g f\,(a,b) = W_{e^{jx}} f\left(\frac{1}{\omega}, 0\right)$$

$$= \int f(x)\, e^{-j\omega x}\, dx = F(\omega)$$

which is a Fourier transform. Rigorously, several mathematical difficulties arise with this substitution, but the intuitive interpretation and inverse relationship between frequency and scale are the desired results.

This interpretation of the Fourier transform as a wavelet transform is exploited throughout this book. Note that the translation parameter, b, is always zero (thus reducing from two-dimensions to one-dimension). This makes sense because the exponential mother wavelet cannot be "local" in the translation parameter (it looks the same for any translations that are integer multiples of the cycle period of the exponential). For mother wavelets that are local in x, the translation parameter will not be identically zero and, thus, this parameter provides valuable information regarding the energy distribution of the signal f.

For wavelet transforms observe that all of the scaled and translated mother wavelets (Figure 1.4 and Figure 1.5) have the exact same number of cycles - the mother wavelet, g, is simply expanded or compressed and slid along the axis. If these scaled and translated mother wavelets are considered as a window (a function that tapers in both positive and negative directions), then this window size will change for each different scale. This changing window size causes the number of cycles or oscillations *within a fixed interval* to change too. If the scale is less than one, then the mother wavelet or window is compressed and more cycles of the mother wavelet will be included in a shorter time interval. More cycles in a shorter interval implies higher frequencies. Thus, the wavelet transform naturally maps a smaller time interval for higher frequencies and a longer time interval for lower frequencies. This frequency/window size interaction is straight forward for simple exponential mother wavelets; however, since mother wavelets are not limited to be simple exponentials and can be very complicated functions, this interpretation must be generalized. Chapter 3 discusses more general resolution properties of wavelet transforms and their relationship to the mother wavelet.

As a physical analogy of a wavelet transform, consider star gazing with a telescope. The lens of the telescope acts as the mother wavelet and the sky as the input signal. When the telescope is out of focus so as to view the entire sky, the telescope does not need to be moved at all to characterize all of the sky at this coarse "focus." This smeared view cannot resolve closely spaced stars or planets. As the telescope is focused, so as to examine a smaller region of the sky, it must be moved (or translated) several times to view the whole sky. When the telescope is focused (scaled) to its finest resolution, then it must be moved many times to view the whole sky at this fine focus.

The analogy is that the sky (the input, f) is to be represented by multiple views (wavelet coefficients) through a lens (mother wavelet) at different scales (focuses or resolutions), and different "look" directions (translations). The movement or swinging of the telescope is the translations, b. The focus of the telescope is the scale, a. At each scale and translation (focus and position) a new "look" or view (wavelet coefficient) of the sky is created. The lens acts as the mother wavelet. Although the lens size has not changed, the response of the entire telescope, due to the focusing, appears to be another lens that has changed in size. The lens properties are identical at each focus or scale - just like the scaled and translated mother wavelet. The set of all views can be combined to reconstruct a "multi-resolution" picture of the sky. Further theoretical details of multiresolution wavelet decompositions can be found in [Dau, Mal, Mey, Rio] and several chapters of this book.

Unlike some other transforms, the wavelet transform does not have meaning unless another function is associated with it, the mother wavelet function, g. One signal, f, has infinitely many "wavelet transforms" because many different mother wavelets exist (for each different mother wavelet, a different wavelet transform exists). For the Fourier transform only the exponential kernel is allowed and the transform becomes unique without specifying the kernel. When a particular mother wavelet is specified, then a signal has only one wavelet transform. The wavelet domain extends over the scale and translation parameters, "a" and "b," respectively. The wavelet coefficients exist only over scale and translation parameters not the original independent variable (x in equation (1.2)). The independent variable being scaled can be time, space or any appropriate parameter, and x is simply a representative parameter.

Consider another example of a more physical time scaling operation. Figure 1.7 shows a rider on a motorcycle yelling to a boy who is standing still. Assume that the motorcycle moves at a constant velocity directly at the boy. The speech process that is formed by the boy will be a time compressed version of the speech process heard by the rider. Since the rider forms the speech, the speech can be understood by the rider but not by the boy. The time compression forms a modified speech signal. Even if the boy could make sense of the signal, the "standard," non-scaling speech recognizers may perform poorly. The time scaling operation is impossible to model with the commonly employed "convolutional" system models (Chapter 5).

Example Requiring Wideband Processing

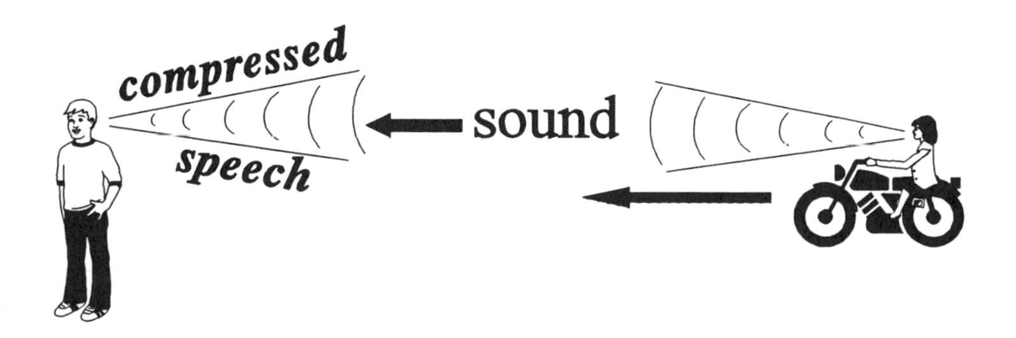

Figure 1.7: Time compressed Speech Signal Example

Now consider the boy yelling and the speech reflecting from the motorcycle and rider. The speech process would return to the boy but it would be compressed just as if the rider had shouted it as in the previous example. The first order model of a general reflection process is a linear time warping - a compression or dilation (see any standard physics text or [Kel]). When signals can be further limited or constrained to be narrowband (a signal with a small bandwidth compared to the its center frequency), then this time scaling can be approximated by frequency (or Doppler) shifting. For the speech signal (which is *not* a narrowband signal) and almost any motorcycle speed, the difference between the original voice signal and the received signal cannot be properly modelled by a frequency shift; time scaling would be necessary. If frequency shifting is employed instead of time scaling then the processing could have less gain and/or resolution and become more sensitive to noise. Thus, for this example and many applications, a more general wideband or wavelet model is required to more appropriately model the reflection process. This would be a system model because no mother wavelet is required to represent the reflection operation.

For the compression in the previous example the scale is positive - the signal was not reversed in time. However, the scale can be negative. Scaling of the independent variable by any negative number reverses the independent variable axis. The signal is backward in time. A scale of -1 causes a reflection around 0 on the independent variable's axis (compare Figure 1.1 to Figure 1.5); the reflection corresponds to running your VCR in reverse at play speed (watching the moving backwards). If a function is symmetric around 0 then a negative scale has no effect on the signal. For the wavelet analysis performed in this book *both positive and negative scales will be allowed.* Some researchers only use positive scales but for generality and notational simplicity this book allows the scale to be any nonzero real number.

As previously stated, wavelet theory is a general mathematical tool primarily used for representing things more efficiently. Although the "things" could be the solutions to differential equations or many other abstractions, this book concentrates on the utilization of wavelet theory for "signals and systems" that abstractly model some physical phenomenon. The field involving signals and systems is often termed signal processing.

Signal processing includes the following operations: converting some physical phenomenon (i.e., reflected light, a contracting muscle, air motion) into a "signal" (or vice-versa), conditioning and manipulating the signal(s) to extract or encode desired information, and interpreting and/or reacting to the extracted information. The signals can be acoustic and/or electromagnetic waves, the DOW industrial index, sequences of numbers, some frames of the Godfather movie, the flow of blood in an artery, etc. Wavelet theory is a new mathematical tool that can be utilized to extract or encode the desired information or features.

The applications of wavelet theory in signal processing extend from speech analysis, medical imaging, theoretical mathematics and physics, data compression, communications, oil exploration and seismic sensing, sonar, weather forecasting, stock market modelling, etc. (fields involving trend analysis or patterns over almost any interval of space and/or time). The primary, and most advantageous, application areas are those that have, generate, or process wideband signals (wideband signals are signals that can simultaneously have both short term characteristics or trends and relatively long term characteristics as well - the duration of the interesting portion of the signal is not known a priori). In addition, systems involving time and or space variation (systems that are not time or space invariant as well) are appropriately and efficiently represented by wavelet theory.

Let a speech signal be the original "object." Consider two different speakers, each one with a different accent and one that speaks rapidly and the other that speaks slowly. The process of recognizing the speaker's "message" from a particular sentence usually begins by recognizing individual words (the pieces or the sentence). The words uttered by the two speakers are the same but they occur at different rates and are interrelated differently due to the accents. In the recognition process the listener must account for the different rates by performing a *time scaling* on the different words (time scaled pieces). Even if the entire sentence is not initially understood by the listener, several words may be recognized. This usually ends the breaking down or analysis stage. By using context, past experiences with this accent, and any other available tools, the disconnected but recognizable words can often be used to reconstruct or synthesize the original sentence or message. Thus, the process of breaking "objects" into component pieces, operating on them and reconstructing a new object is a natural process. For this example time scaling the individual pieces (words) was also natural.

Note an important distinction between the jigsaw puzzle example and the speech example; the component pieces of the speech are often very overlapped and connected - some pieces are not easily separated. In addition a lot of redundancy may exist in communicating a message. The redundancy allows incomplete recognition to reconstruct the original message. Many wavelet representations lead to redundant representations. Another distinction between jigsaw puzzle pieces and the speech pieces is the scaling operation that acts on the sections of speech.

Reconsider this sentence recognition problem as a speaker recognition problem. In some speech recognition systems the time scale parameter is a feature used for speaker identification. Having both speakers say the same sentence provides a "reference" input. Each speaker then forms that reference input into speech process. Now by examining the two speech processes can the speakers be identified or recognized? Now the individual words and the message do not matter, only the difference between the two complete speech processes. This speaker recognition problem requires a different approach than the message or sentence recognition problem. The speaker recognition problem concentrates on the differences between the two entire speech signals while the message recognition problem concentrates on identifying pieces. This book considers both of these problems with more general formulations. The "standard" wavelet analysis decomposes a signal into component pieces and operate on these pieces (or their coefficients); this technique is applied in image analysis, coding, speech recognition, etc. Since the pieces are always a scaled and translated mother wavelets, the representation is highly sensitive to the chosen mother wavelet. A wavelet representation of a signal is a distribution across scale and translation relative to a particular mother wavelet (if the mother wavelet is changed then the same signal will be represented by a different distribution across scale and translation - this is detailed further in Chapter 2).

Since reflection from a moving object is considered throughout this book, the mapping from the measured parameters of scale and translation to range and velocity of the reflector is quickly presented. Details and justification of this mapping can be found in the references [Kel, Van]. If a known signal is transmitted, then some time later (this delay is the time translation that will be estimated by a wavelet transform), the reflected signal is received. This estimated delay is mapped to range by simply multiplying it by the speed of wavefront travel in the medium (speed of light, sound, etc.) and dividing it by two to account for the round trip travel, $range \approx (wavefront\_speed \cdot delay)/2$. Similarly, the time scale can be estimated from a wavelet transform of the received signal and this estimated scale maps to the radial velocity of the reflector (the "line-of-site" velocity between the transmitter/receiver and the reflector) as:

$$radial\_vel \approx (wavefront\_speed \cdot (1-scale))/2 = \frac{c}{2}(1-scale)$$

These equations provide one physical relationship for the translation (range) and scale (velocity) parameters.

The latter part of this book addresses the problem of modelling a system or the difference between two systems (each speaker could be modelled as a system with the input being a sentence and the output being the speech process). For this problem the input signal becomes the mother wavelet and *the system model is a distribution over scale and translation. Thus the system model is truly independent of any mother wavelet; each different scale translation distribution represents a different system. System modelling is very different from signal modelling due to the elimination of the mother wavelet.* This book addresses both signal and *system modelling with wavelet theory.*

This book presents new system models that can accurately model systems that quickly change over time and/or space, avoiding the highly constraining stationarity assumption [Van1] and be robustly estimated (unlike some previous time-varying system models). This new system also processes both nonstationary and/or wideband input and output signals.

Again, the application of any representation (from the decimal or binary number representation to wavelet transform representations) depends upon its impact on improving the efficiency of the application. If a discernable efficiency or gain improvement is not achieved, then the representation is not useful for that application. Several signal representations exist, most notably the Fourier representation, that improve the efficiency of many applications. The Fourier representation is advantageous in applications that have narrowband (cyclic) behavior consistent across time or space (or statistically stationary). For some applications an efficient signal representation may not exist and the wavelet and/or Fourier decompositions should not be used; the signals should remain in their original domain (time, space, etc.).

One of the primary goals of this book is to identify and justify the application areas where wavelet theory can be most beneficially utilized. Besides efficient representations for specific types of signals, wavelet theory also provides an efficient representation for the "actions" of specific types of systems. These signals and systems are identified and discussed.

Several biological functions, such as hearing and vision, appear to operate on signal information by weighting each octave nearly equally rather than proportional to its bandwidth as information theory suggests [Mal3]. The octave band weighting is natural for the wavelet transforms (these transforms are referred to as dyadic wavelet transforms) and this relationship has been exploited in several initial wavelet applications. Besides modelling signals, systems must be modelled that operate on these wideband multiple octave signals. These systems will not necessarily operate on signals by moving them to the next entire octave. Instead the systems/channels may perform very subtle scale changes on these signals. Wideband interferometry would be an example of a system that could not possibly be modeled by octave band shifts but would require fine scale changes.

It is important to note that the objectives of wideband-nonstationary *signal representations* are often very *different* than the objectives of time-space-varying *system representations*. Both the signal and the system representations exploit the wavelet theory features but the details of the resolution requirements are often significantly different; these differences and the appropriate operators for each one are detailed in this book. The applications and further theoretical extensions in this book concentrate on formulating signals and system operations with wavelet theory.

"Wideband signals" directly exploit the additional capabilities of wavelet theory as does space and/or time varying (STV) systems. For some applications, motion and spatial gradients can be construed as the same actions. An example of a physical problem involving motion is the general reflection process (not simply a Doppler or frequency shifting operation) [Kel, Van2]. Thus the primary applications addressed in this book are related to wideband signals and time and/or space varying systems *and their combination: wideband-nonstationary signals passing through space-time-varying (STV) systems*. These applications are quite general and extend (in many cases replace) the current theory (time-varying impulse responses, multi-frequency representations, etc.). Image processing applications - images are wideband signals - are included for completeness but are only briefly covered due to the extensive coverage in the many stated references.

Recent advances in wavelet theory have concentrated on efficient representations with orthogonal wavelet bases. These bases are a significant breakthrough and have already had significant impact in image processing, theoretical mathematics, and other areas. However, without down-playing the vital significance of the orthogonal wavelets, efficient representations can also be achieved with nonorthogonal wavelet transforms.

The requirements of the wavelet transform representation change for each different application. The most notable change between the image processing application (biorthogonal, orthogonal and multiresolution wavelets) and the system or channel characterization application (nonorthogonal wavelets) is typically the resolution requirements in the scale parameter; scales that change by a factor of 2 may be acceptable in image analysis, while scales that change even by a factor of 1.00001 may not be adequate for system or channel characterization. These different requirements on the wavelet transform itself limit the particular transforms that can be employed in a specific application. Thus, not only are wavelet transforms presented, but different types of transforms are fit to appropriate application areas. The characteristics of the mother wavelets are thoroughly investigated.

In environments that have motion, many of the present sensing or imaging systems employ short pulses so that the motion in the environment is negligible over the signal duration - creating a "stationary" scattering process (non-changing over that short observation or processing interval). However, due to these short pulses the frequency or scale duration is very poor (typically the scale or frequency resolution is inversely proportional to the signal duration). In addition, the large frequency extent (due to the short pulses) and any motion in the environment will often lead to aliasing.

The poor resolution in scale or frequency is often ignored because only the pulse's delay (range) is exploited in further processing structures or estimators. However, if a simultaneously good measurement of scale or frequency and delay could be obtained, then surely this additional information should significantly improve the processing or estimators (the gains and/or resolutions). Essentially, the simultaneous measurement of scale (frequency) and delay provides a two-dimensional measurement that is much "richer" than the one-dimensional, delay measurement. This result has been stated and verified in many references, but Chestnut [Che] provides the most direct statement and application of this effect for the passive sensing application.

The implementation for simultaneously achieving fine resolution in delay and scale necessarily involves high time-bandwidth product signals. These large time-bandwidth product signals and the systems that act on them are well modelled by wavelet transforms.

Technical Specifics and the Research Direction of this book

This book demonstrates the application of wavelet theory to a diverse set of *general* applications including systems theory, scattering theory, signal design and many other fields. By formulating these general applications with wavelet theory, wavelet theory can be applied in many of the more detailed applications. For example, the general systems and scattering theory reformulated with wavelet concepts can be applied to noninvasively image blood motion within a human body, or to model the human speech process. Thus, several general "applications" have been improved but each of these general applications may be employed in a huge set of "specific" application areas.

One general extension to wavelet theory presented in this book is a new operator, the Mother Mapper Operator. The Mother Mapper Operator efficiently creates multiple mother wavelet representations of a single function. Several applications are presented, but the primary thrust of this book is the detailed application of wavelet theory to space and/or time varying (STV) systems. The characterization of the STV systems exploits the time scaling operation of wavelet theory. However, the STV Wavelet Operator collapses to the linear time invariant (LTI) system model when the system does not involve any motion or non-unity scaling (the STV wavelet operator becomes the convolution operator and the 2-D wavelet characterization becomes the 1-D impulse response). Thus the new STV Wavelet Operator is a generalization of the LTI system model for systems that process wideband (multi-octave) signals that may have significant motion in them. The STV wavelet representation of a system is a two-dimensional distribution over scale and translation; whereas, the LTI system representation is the one-dimensional impulse response. If the STV representation is zero everywhere across the two-dimensional plane except along the unity scale line (scale = 1), then this particular STV representation can be interpreted as an impulse response.

After establishing the mathematical theory, the STV system characterization is formulated in a physical environment. In a physical environment the pure mathematics is not sufficient to describe the utility of the wavelet processing. By including the physics of the problem, the ambiguity of the processor and its reference frame are resolved. The wideband ambiguity function is, essentially, a wavelet transform and these relationships are detailed. Wideband ambiguity functions and wideband scattering are addressed and physically interpreted for both active and passive sensing systems.

Much of the initial wavelet applications concentrated exclusively on time-frequency distributions or image analysis. Due to the potential benefits and profitability of image analysis, most of the wavelet research continues in this application. For image analysis, the wavelet theory efficiencies center around the "multiresolution" and "orthogonal or biorthogonal" wavelets (these terms are discussed in Chapter 2) and, thus, most of the research is also related to multiresolution and orthogonal wavelets. These wavelets and their desirable properties are presented and discussed along with the image analysis application. Several problems with these wavelet representation are identified and discussed as well.

New wavelet operators are derived (without requiring advanced mathematics) to improve the resolution and efficiencies associated with the nonorthogonal (general) wavelet theory. Providing simple mathematics does limit the generality and elegance of some of the results; the more advanced reader can refer to the research literature for the mathematical details and generalities. The new operators allow *multiple mother wavelet representations* to be efficiently considered rather than being constrained to a single mother wavelet. The application of the new operators and the general wavelet theory to *wideband, nonstationary, signal processing, and space-time-varying system modelling* is detailed with an emphasis on the *physical interpretation.*

Appendix 1-A: Time Referencing

When time scaling is involved a time reference is required. Space varying systems would similarly require a spatial reference but only time scaling is discussed here. The affine operation is a combination of both scaling and translation. Reconsider the movie example that was discussed in this chapter. Assume that the movie is 2 hours long and was originally scheduled to be shown at "2" at play speed. The movie would have ended at 4. If instead the movie is delayed by 3 hours and will be played at twice "play" speed (fast forwarded), then the movie would start at 5 and only run for 1 hour (instead of 2). Thus, the delayed and time compressed movie would start at 5 and run until 6.

Consider the affine operation and the scale and translation (delay) parameters. Let the movie be denoted by the function g(t). Time, t, is absolute time that keeps going forever. Allow the movie to be fixed in time so that at t=2 the movie starts or g(2) is the first frame of the movie. In addition, since the movie is 2 hours long, g(4) is the last frame of the movie.

To create a delayed and time compressed version of this movie its time argument must be warped and translated so that the "new" movie is:

$$g(t') = g\left(\frac{t-b}{a}\right)$$

where t is absolute time. For a 2 hour movie that is fast forwarded by a factor of 2, only 1 hour of absolute time should pass by. Thus, for the whole movie, t should only change by 1 where as t' must change by 2 hours (to get from the first frame to the last frame the argument of "g" must change from 2 to 4). Similarly, if the movie is delayed by 3 hours of absolute time, then this maps to some delay or translation, b, in the warped time.

The goal is to map absolute time such that the movie is shown 3 hours later (in absolute time) and twice as fast (fast forward). An affine operation can be used to map absolute time to the desired warped "movie time." An affine operation can act on the absolute time, t, to warp it to a new movie time t'. Remember that the movie starts at an argument of 2 and ends at an argument of 4 (assume $g(arg) = 0 \text{ for all } arg<2 \text{ or } arg>4$). What is the scale, a, and the delay or translation, b, that will map the absolute time into the new "movie" time?

The new argument of g is t' and absolute time, t, is mapped to t' by a and b. Since the movie is fast forwarded by a factor of 2, then the scale parameter, a, must be one half (a=0.5). This scale will cause 1 hour of absolute time to become 2 hours of "movie" time. Now consider the translation, b. Since the first frame of the movie starts when its argument is 2, then t' must be 2 at the start of the movie. However, the movie doesn't start in absolute time until 5 (t=5) due to the 3 hour delay. Thus, when t=5, $t'=2$. Since a=0.5, solving $t' = \frac{t-b}{a}$ yields b=4 and so $t' = \frac{t-4}{0.5}$. The absolute time, t=5, is mapped to $t'=2$ so that the start of the movie is delayed 3 hours in absolute time and the movie ends at t=6

(t'=4) so that the movie only lasted an hour of absolute time (but 2 hours of movie time). The proper scale and translation parameters to accomplish this warped movie time are a=0.5 and b=4.

Unfortunately, this same scale and translation will not work for the movie that was originally to be shown at noon (t=0) and is also supposed to be delayed by 3 hours and fast forwarded by a factor of 2. Let this noon-time movie be $g_2(t)$. Its first frame, $g_2(0)$, occurs when its argument is 0 (noon). Let its new argument be t'. Then, due to the 3 hour delay, this movie should start at t=3. Inserting t=3 into $t' = \frac{t-4}{0.5}$ yields t'=-2, not t'=0 that is required for the first frame of the movie. The same scale and translation parameters do not work for this movie that worked for the last movie. To properly map t=3 to t'=0 requires b=3. The only difference between the two movies is their original starting time or *time reference*. Both movies are 2 hours long and are both delayed by 3 hours and fast forwarded by a factor of 2. Why can't the same affine operation acting on absolute time do the same thing to both movies? Because the effect of the time scaling changes depending upon when it is applied - time scaling is a time-varying operation. This will be further detailed in later chapters, especially Chapter 5, but its understanding is critical for any applications that apply time scaling in their models.

Before deferring a further discussion to later chapters, suppose that the two movies are concatenated to form one, 4 hour long documentary instead of two separate movies. What is the proper affine operation to act on absolute time that would cause this documentary to be delayed by 3 hours and be shown at fast forward speed? Since the documentary was originally scheduled to start at noon or t=0 (which is the chosen time reference) then the proper affine operation for the entire documentary would be a=0.5 and b=3. This affine mapping will work for the entire documentary - including the second half which was the first whole movie!

Besides this specific case concerning movies, consider a general system that performs time scaling. If the input to the system is x(t), then the output will be y(t)=x(at). If the input to this same system is x(t-d), then the output will be x(at-d). The system delays a signal by differing amounts depending upon when the input was applied - the response of the system is dependent upon time, the system is time varying. Please refer to Chapter 5 for further details regarding time-varying models and time scaling.

Chapter 2: The Wavelet Transform

Chapter 1 presented the concepts involved with wavelet theory, especially the scaling operation, and avoided all of the mathematical rigor. This section supports the conceptual statements by providing the mathematical justification. General, continuous-time wavelet transforms are initially discussed. Then the "discrete" wavelet transform is presented. The resolution properties of wavelet transforms are touched upon but then deferred until after ambiguity functions are discussed. For the less mathematically inclined reader the proofs can be avoided without losing continuity.

Wavelet Transform Definitions and Operators

Before defining the wavelet transform, admissible functions are defined. For a function to be a mother wavelet it must be admissible. Recall, from the discussion in Chapter 1, that for a function to be a wavelet or mother wavelet it must be oscillatory and have fast decay toward zero. If these conditions are combined with the condition that the wavelet must also integrate to zero (its "d.c." or zero frequency component is zero), then these three conditions are the "non-rigorous" admissibility condition that must be satisfied for a function to be a mother wavelet. Essentially, admissible functions are bandpass signals - these signals cannot have zero frequency components and they must decay, so they will not have infinite frequency components either. Note that most signals that travel through a medium or in free-space are finite duration, bandpass waves, so this requirement is not very restrictive.

Although a Laplace transform also includes a kernel function (a decaying exponential) that both decays and oscillates, the decay is always centered around zero (unlike these wavelets), frequency shifts are used (instead of time scaling), and the kernel of the Laplace transform is an exponential (exclusively). As will be emphasized throughout this book, the mother wavelet (kernel of the wavelet transform) can be almost any function.

More rigorously, an $L^2(\mathbf{R})$ function (a finite energy function - square integrable over the range of its independent variable), g, which can be either real or complex, is an **admissible function** if:

$$c_g = \int_{-\infty}^{\infty} \frac{|G(\omega)|^2}{|\omega|} d\omega < \infty \tag{2.1}$$

where $G(\omega)$ is the Fourier transform of g. Note that the lower limit of integration is minus infinity instead of 0. This is required if the mother wavelet is complex and has a spectrum that is nonsymmetric about zero frequency. This admissibility condition is sufficient (may be more restrictive than required) but is not necessary (some functions are admissible but do not satisfy this condition). A more general necessary and sufficient condition (the complete definition of admissibility) for functions to be admissible is defined with group theoretic concepts in [Gro2, Hei]. In summary, admissible functions (and mother wavelets) are those that cycle (oscillate), have finite energy, and have an average value of zero. Most natural signals satisfy these properties; energy usually travels as wave packets (oscillates and has an average value of zero - its average square value is the energy). Thus, most natural signals would classify as admissible functions.

The **wavelet transform** operator, $\boldsymbol{W}_g$, maps a finite energy or $L^2(\mathbf{R})$ signal that is real or complex valued as follows: $\boldsymbol{W}_g : L^2(\mathbf{R}) \rightarrow L^2(\mathbf{R}\setminus\{0\} \times \mathbf{R})$. Stated less mathematically, any finite energy signal is mapped from the time or space domain to a finite energy two-dimensional distribution in the scale-translation or *wavelet domain*. The **wavelet transform** of a function, f, with respect to a given admissible mother wavelet, g, is defined as:

$$\text{wavelet domain coeff at scale } a \text{ and translation } b$$

$$= \boldsymbol{W}_g f(a,b) = |a|^{-\frac{1}{2}} \int f(x)\, g^*\left(\frac{x-b}{a}\right) dx \tag{2.2}$$

$$= \left\langle f, \frac{1}{\sqrt{|a|}} g\left(\frac{x-b}{a}\right)\right\rangle = \langle f, g_{a,b} \rangle = \langle f, U(a,b)\, g \rangle$$

where superscript "*" denotes complex conjugate and $\langle \cdot, \cdot \rangle$ is an inner product (shorthand notation for the correlation integral defined in this equation). Refer back to Figure 1.6 for an example of the wavelet transform's operation. Note that this definition requires $g(x)$ to be an admissible function. The wavelet element, $g_{a,b}$, is defined by a **unitary affine mapping** $U(a,b) : g(x) \mapsto \frac{1}{\sqrt{|a|}} g\left(\frac{x-b}{a}\right)$ or, less mathematically, $g_{a,b}$ is a version of the mother wavelet, $g(x)$, that has been scaled by the scale parameter, a, and translated by the translation parameter, b. The scale can be any nonzero real number, $\mathbf{R}\setminus\{0\}$; refer back to Figure 1.5 to see a negatively scaled mother wavelet. The "unitary" refers to the energy normalization performed by the $\frac{1}{\sqrt{|a|}}$ term that keeps the energy of the scaled mother wavelet equal to the energy of the original mother wavelet. The mother wavelet can be normalized further to have an admissibility constant, c_g, that is unity. A unity admissibility constant allows the constant to be dropped and yield an isometric wavelet transform (energy is the same in both the x domain (time or space) and the wavelet domain (scale and translation)). This allows any wavelet coefficient at any (a,b) to be "fairly" compared or combined even across a large range of scales. Note that for continuous wavelet transforms the choice of the mother wavelet is only constrained by the admissibility condition. One is free to choose the mother wavelet for optimal behavior in the particular application of interest.

By examining how a weighted set of wavelets approximate a function, the structure of the inverse transform can be established from the "forward" wavelet transform. A set of wavelets generated from one particular mother wavelet can be used to construct approximations of a function (a wavelet domain representation of that function). A function is approximated by a weighted sum (or integral) over the wavelet set (scaled and translated mother wavelets). Each weighted wavelet acts as a building block and when all the blocks are summed together, an approximation is formed. To obtain the appropriate weight for each element of the wavelet set, the function being approximated is projected onto each element in the wavelet set; the result of each projection is a scalar number (real or complex) called a wavelet coefficient of the function f with respect to the mother wavelet g, or just $\mathbf{W}_g f(a,b)$. This approximation action is the wavelet transform. See Figure 2.1 (discussed subsequently).

Besides being defined in equation (2.2), the projection process is a correlation (a measure of how well two signals match). The signals that are being "matched" or compared are the original function, f, and scaled and translated versions of the mother wavelet, $g\left(\frac{x-b}{a}\right)$. In Figure 2.1 the projection, correlation, or inner product is denoted by an "I" with a circle around it to indicate an "inner" product. The inner product operator is expanded in the lower right corner of the figure to demonstrate its structure (the inner product should strictly include the conjugation operation too but the conjugation is shown to act on the wavelets to demonstrate a difference between the wavelet transform and the subsequently described inverse wavelet transform). A wavelet domain coefficient is computed for each particular scale and translation (a,b) value. This coefficient is denoted, $\mathbf{W}_g f(a,b)$. Determining these coefficients, and thus the functional approximation, is the wavelet transform.

The **wavelet transform** defined in equation (2.2) is an analysis filter. It breaks a signal down into component pieces and these pieces are examined or operated on instead of the original function. The inverse wavelet transform or reconstruction filter is a synthesis filter. The synthesis filter puts the pieces back together again (correctly). Why break a signal apart and then put it back together? Because significant insights, gains and efficiencies can be obtained through the analysis process and operating on the pieces rather than the original function. Many entire subjects are built around this concept (mathematical analysis, principle component analysis, eigen-analysis, coding theory, discretization, Fourier analysis, etc.) but a simple motivational example might help.

Consider the sounds of an orchestra. The composite sound is a mix of many instruments and is a complete sound signal. However, a good listener can distinguish or extract the sounds emanating from a single particular instrument; a "good ear" can *analyze* the component pieces of the composite sound signal. If the

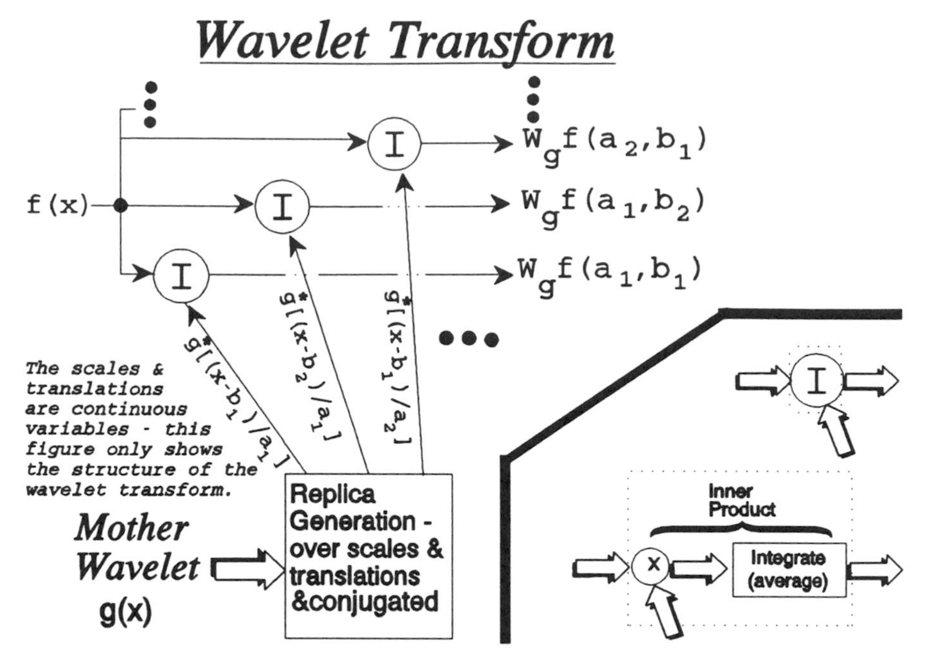

Figure 2.1: Wavelet Transform Structure

listener can also compose and transcribe this music, then a concise or efficient representation of the (majority) of the sound can be formed (a musical score). If it was desired to hear this same sound at a different location, then, in terms of information to be transmitted, it would be much more efficient to transmit the musical score than a recording of the music itself. Thus, a data compression or efficient representation has been achieved by analyzing the signal's component pieces. A new representation is created from only the pieces that are the most significant (energy is one measure of significance). Obviously, some loss in quality may be incurred and the gains in efficiencies must be traded off against distortion and other losses; each application has its own set of tradeoffs. Some researchers have proposed wavelet transforms for analyzing musical sounds and creating efficient representations [Com]. Before considering further applications of wavelet theory, several specific wavelet transform are computed and their properties are examined.

Wavelet Transform Examples

The wavelet transforms of several elementary functions are presented in the next four figures to display the characteristics of the wavelet domain representation. Note the increase in dimensionality of the wavelet representation; the dimensionality increases by a factor of two. For the first three wavelet transforms the mother wavelet was a complex Morlet mother wavelet (or a Gaussian weighted tone) with about six significant cycles in it (the imaginary part of this complex mother wavelet was shown in Figure 1.1). For these figures a 30 dB range of magnitude is displayed - if the magnitude was less than 30 dB below the peak, it was set to zero along with its corresponding phase. These figures are not studied in detail until the wavelet transform properties are examined; however, note the time localization property of the wavelet transform for the impulse (or delta function) and the edges of the rectangle. Frequency or scale localization can also be observed from the wavelet transform of a tone (sinusoid).

Several important conclusions can be made from these figures. First, note that both the magnitude and the phase localize in both time and scale (frequency). The localization in phase is observed from the converging phase ridges (not the zeroing due to small magnitudes). The simultaneous localization is due to the entire envelope being moved; the mother wavelet is translated. For the impulse or delta function, the wavelet transform will simply be the mother wavelet with translation replacing the time parameter (look at equation (2.2) with the input being an impulse - $W_g\,[\delta(x)]\,(a,b) = g\left(\frac{-b}{a}\right)/\sqrt{|a|}$). At each different scale the wavelet distribution is a scaled mother wavelet.

Note that the phase of the wavelet transform and the real part of the wavelet transform are essentially redundant information (and, thus, look alike). To be consistent with matched filter or correlation processing, the phase will be chosen over the real part for the representations in this book.

For all of these cases the mother wavelet was complex. For both the impulse and the rectangle, the input was real. If a real mother wavelet was used (the real part of the complex mother wavelet), then only the real part of the wavelet transform would be obtained. The real part of the wavelet transform can be very poor for identification/detection. If the wavelet transform is only evaluated at points that are near the nulls of the real part of the transform, then those evaluated points can be sensitive to noise or might suggest that the signal is not present when it really is present. These nulls in the real part can lead to incorrect results and invalid conclusions. This is similar to matched filter processing with real signals as opposed to complex signals; processing with a real signal produces nulls in the matched filter's output and leads to sensitivities and a less robust filter. Usually, for general signal processing it will be desirable to have complex mother wavelets to avoid the possibility of a null (later this condition will be interpreted as a constraint on the density of the wavelet domain "hypothesis grid"). For most wideband signals, only real samples should be processed (as justified in Chapter 3 where the analytic or complex signal should not be formed for some signals). For those wideband signals it is important to maintain the noise immunity by using complex mother wavelets and examining the magnitude of the wavelet transform.

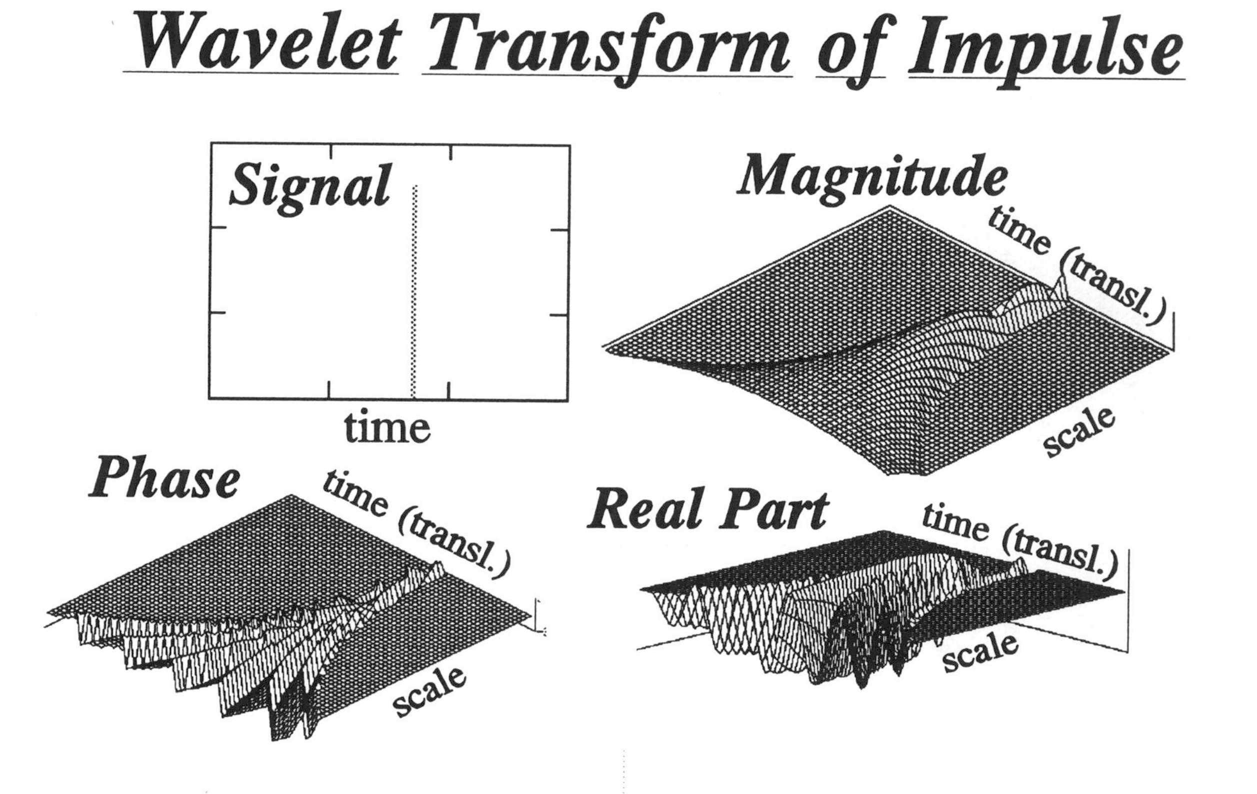

Figure 2.2: Wavelet Transform of Impulse

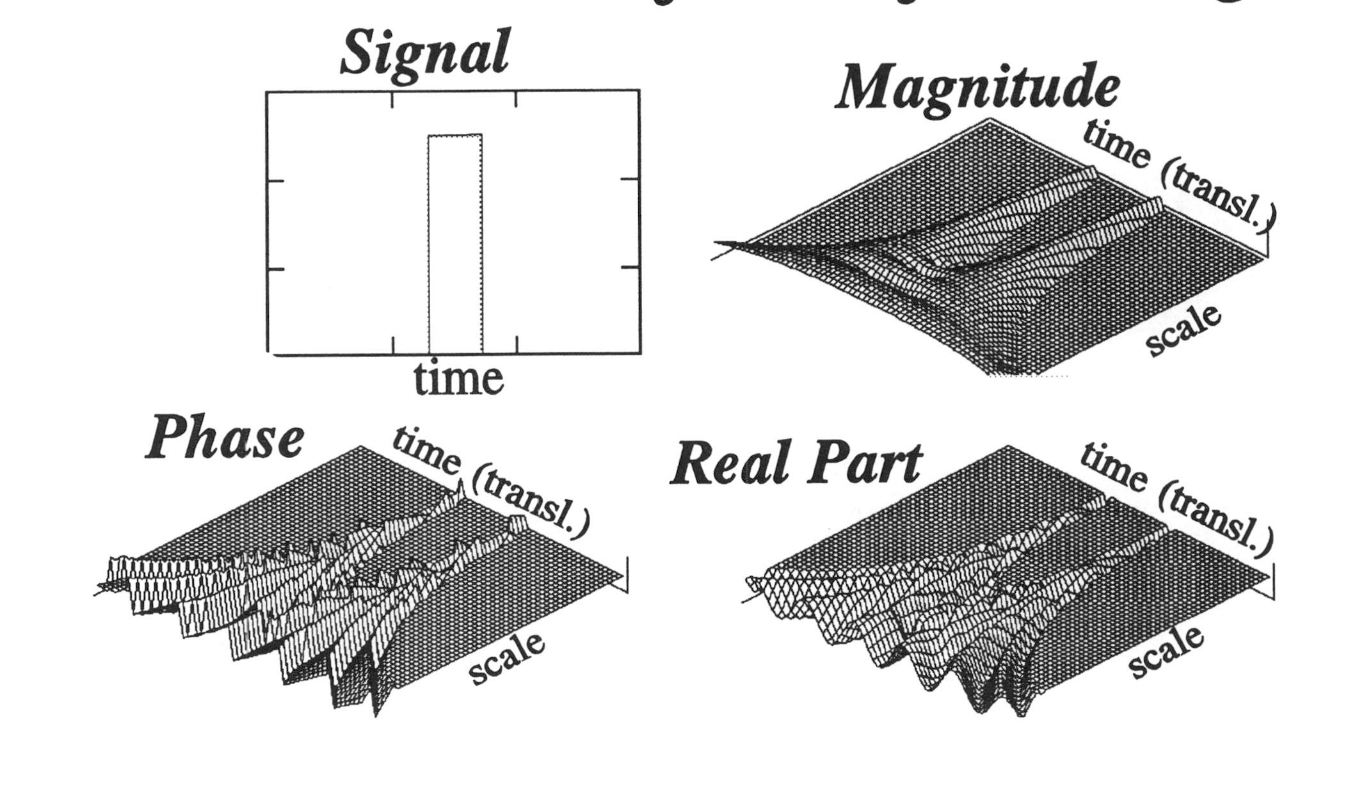

Figure 2.3: Wavelet Transform of Rectangle

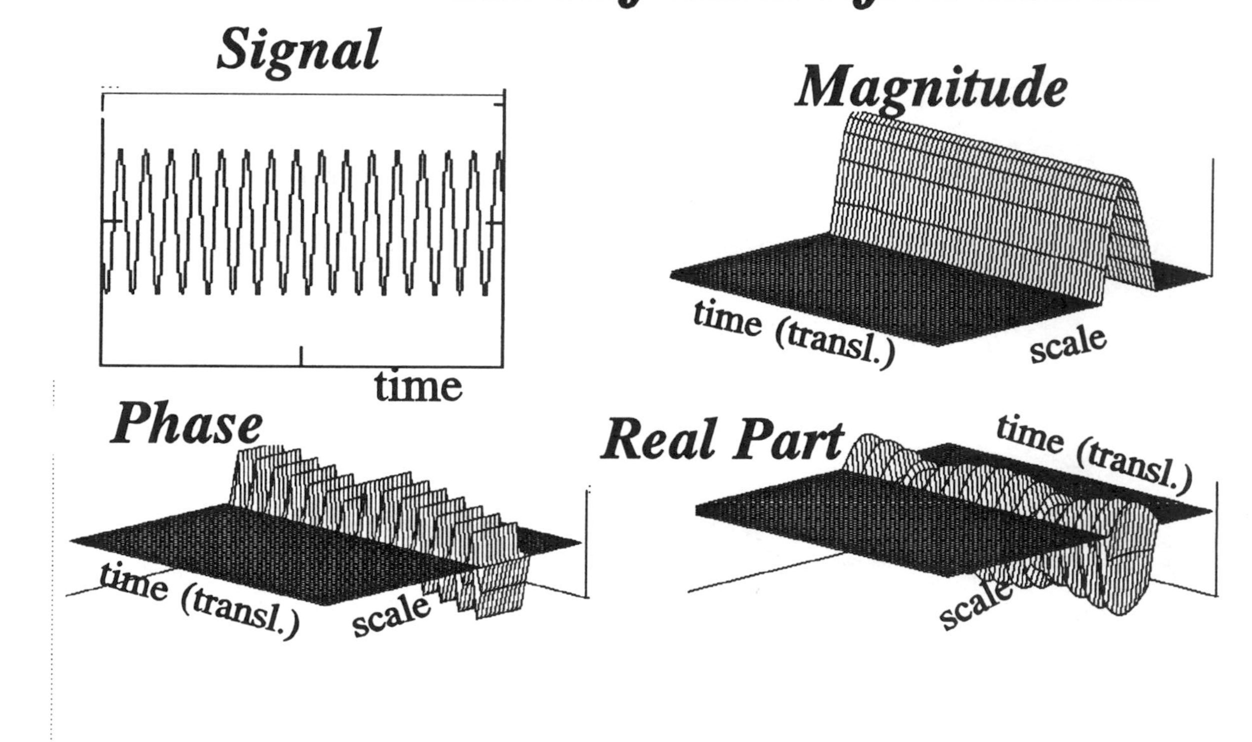

Figure 2.4: Wavelet Transform of a Tone

As mentioned previously, and emphasized throughout the book, the wavelet transform representation and its properties are dictated by the mother wavelet. Because the set of mother wavelets is so large (essentially any bandpass signal) the freedom of choosing a mother wavelet makes general wavelet transform characterizations of a particular function nearly impossible. A valid statement for one mother wavelet can be completely invalid for another mother wavelet. As an example consider the wavelet transform of the same rectangular signal as in Figure 2.3, but now with respect to a FM signal with both quadratic and linear modulation (a sophisticated mother wavelet). This new wavelet transform's magnitude and phase are displayed in Figure 2.5. Both Figure 2.3 and Figure 2.5 are wavelet transforms of the rectangle signal; however, the qualities of the signal in the transform domain are more distinguishable in Figure 2.3 (such as edges of the signal). For general signal qualities represented in the wavelet transform domain, a simple mother wavelet should be employed. For many other applications where gain and resolution are important more complicated or sophisticated mother wavelets might be used.

The characterization of mother wavelets being "bandpass" can be deceiving. A mother wavelet can be wideband and have multiple simultaneous frequencies that are at significantly different frequencies. The Fourier spectrum of a mother wavelet can have many "holes" or frequencies with very little energy in between frequencies that have a lot of energy. The admissible constraint is not a very restrictive condition and the bandpass interpretation should not be accepted too literally.

In addition, wavelet transforms are linear. If a composite signal is created by forming a weighted sum of separate signals, then the wavelet transform of the composite signal is the weighted sum of the wavelet transform of each separate signal. For a further tutorial discussion with more wavelet transform examples and properties refer to the references [Com, Rio]. Having presented some examples of wavelet transforms, the theory of the inverse wavelet transform is investigated next.

The Resolution of Identity

Returning to the wavelet transform theory, an established identity, the **resolution of identity**, is presented. The resolution of identity is utilized in the derivation of the continuous-time inverse wavelet transform, the derivation of the mother mapper operator, and in the study of the wavelet transform resolution properties; thus, the resolution of identity is presented separately. The resolution of identity is the primary mathematical tool that is utilized throughout the rest of the book - it is necessary to understand the resolution of identity to follow the derivations presented in several sections of this book.

Mathematically, the resolution of identity is an isometry (equal energy on each side of the equal sign) that linearly maps a product of two inner products to a single inner product. The resolution of identity [Kla] is a specific case of a more general theorem (the Frobenius-Schur-Godement Theorem in square integrable group representation theory [Gro2]) but only this specific case is required for this book.

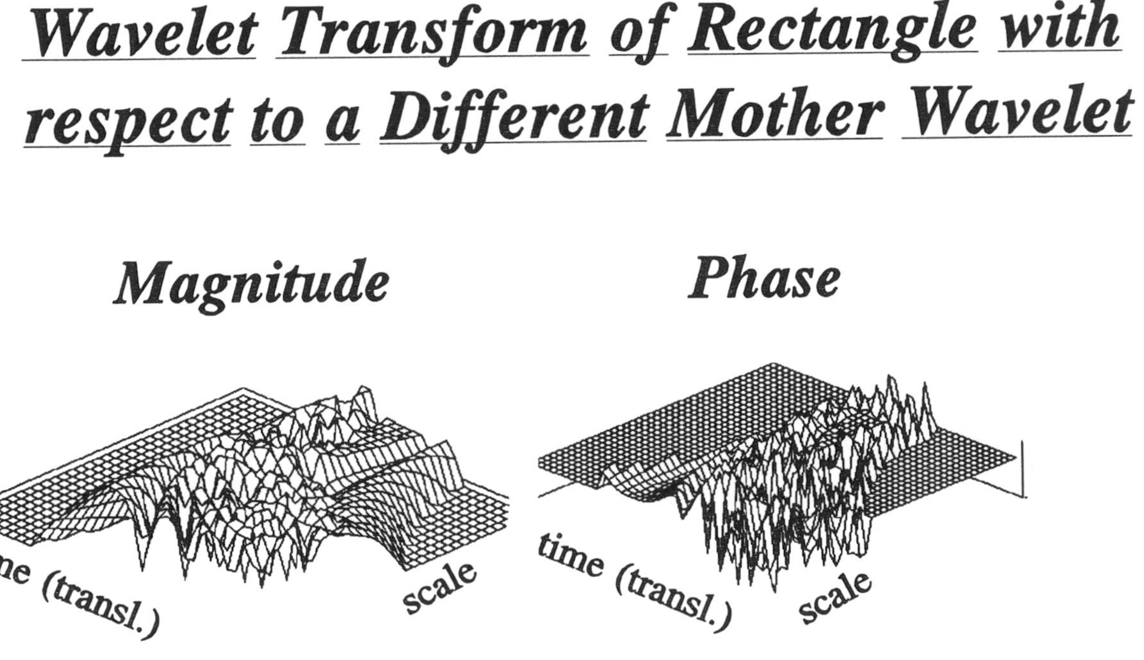

Figure 2.5: Wavelet Transform of Rectangle with respect to a New Mother Wavelet

Practically, the resolution of identity is a formula that allows two wavelet transforms to be combined into one wavelet transform (an operator that maps a product of inner products to a single inner product). The motivation for presenting and deriving this theorem is due to its generality; several more specific formulas can be derived simply by changing the arguments in the resolution of identity. Researchers have used the resolution of identity in the derivation of the continuous inverse wavelet transform or reconstruction formula [Dau2] and this is also done in the next section.

The following proof is suggested in [Dau2]. Since the details of the proof are employed later, the theorem is derived here before its initial application.

<u>Resolution of Identity Theorem</u>: If $f_1(x)$, $f_2(x)$ and $g(x)$ are functions finite energy functions (in $L^2(\mathbf{R})$), $g(x)$ is an admissible function, $g_{a,b}(x) = \frac{1}{\sqrt{|a|}} g\left(\frac{x-b}{a}\right)$, and $(a,b) \in (\mathbf{R} \setminus \{0\} \times \mathbf{R})$ then:

$$\int_{-\infty}^{\infty} \frac{1}{a^2} \int_{-\infty}^{\infty} \langle f_1, g_{a,b} \rangle \langle g_{a,b}, f_2 \rangle \, db\, da = c_g \langle f_1, f_2 \rangle \, . \tag{2.3}$$

<u>Proof:</u> Reformulating the inner products in the Fourier domain:

$$I = \int_{-\infty}^{\infty} \frac{1}{a^2} \int_{-\infty}^{\infty} \langle f_1, g_{a,b} \rangle \langle g_{a,b}, f_2 \rangle \, db\, da \tag{2.4}$$

$$= \int_{-\infty}^{\infty} \frac{1}{a^2} \int_{-\infty}^{\infty} \int_{-\infty}^{\infty} \int_{-\infty}^{\infty} F_1^*(\omega)\, G_{a,b}(\omega)\, G^*_{a,b}(\omega')\, F_2(\omega')\, d\omega'\, d\omega\, db\, da$$

Using the Fourier transform, $\boldsymbol{F}$, and the scaling property that states:

$$G_{a,b}(\omega) = \boldsymbol{F}\left(\frac{1}{\sqrt{|a|}} g\left(\frac{x-b}{a}\right)\right) = \sqrt{|a|}\, e^{jb\omega} G(a\omega)\, ,$$

$$I = \int_{-\infty}^{\infty} \frac{1}{a^2} \int_{-\infty}^{\infty} \int_{-\infty}^{\infty} \int_{-\infty}^{\infty} e^{jb(\omega-\omega')} |a| G(a\omega)\, G^*(a\omega') \cdot F_1^*(\omega)\, F_2(\omega')\, d\omega'\, d\omega\, db\, da$$

and since $\int_{-\infty}^{\infty} e^{jb(\omega-\omega')}\, db = \delta(\omega - \omega')$, then:

$$I = \int_{-\infty}^{\infty} \frac{1}{|a|} \int_{-\infty}^{\infty} \int_{-\infty}^{\infty} \delta(\omega-\omega')\; G(a\omega)\, G^*(a\omega') \cdot F_1^*(\omega)\, F_2(\omega')\, d\omega'\, d\omega\, da$$

$$= \int_{-\infty}^{\infty} \frac{1}{|a|} \int_{-\infty}^{\infty} G(a\omega)\, G^*(a\omega)\, F_1^*(\omega)\, F_2(\omega)\, d\omega\, da$$

Now substitute $\tilde{a} = a\omega$. Then:

$$I = \int_{-\infty}^{\infty} \frac{|\omega|}{|\tilde{a}|} \int_{-\infty}^{\infty} G(\tilde{a})\, G^*(\tilde{a})\, F_1^*(\omega)\, F_2(\omega)\, d\omega\, \frac{d\tilde{a}}{|\omega|}$$

$$I = \int_{-\infty}^{\infty} \frac{1}{|\tilde{a}|} \int_{-\infty}^{\infty} G(\tilde{a})\, G^*(\tilde{a})\, F_1^*(\omega)\, F_2(\omega)\, d\omega\, d\tilde{a}$$

$$I = \int_{-\infty}^{\infty} \frac{1}{|\tilde{a}|} |G(\tilde{a})|^2\, d\tilde{a} \int_{-\infty}^{\infty} F_1^*(\omega)\, F_2(\omega)\, d\omega \tag{2.5}$$

and by applying Parseval's theorem to equation (2.5):

$$I = c_g \langle f_1, f_2 \rangle\ . \tag{2.6}$$

This proves the resolution of identity formula.

Continuous Inverse Wavelet Transform Theorem

An inverse wavelet transform is now stated and derived for the wavelet transform stated in equation (2.2). The composite operation of applying a transform and then an inverse transform is simply the identity operation - the original function is returned. A visual example of the inverse wavelet transform's operation is shown in Figure 2.6.

This derivation is not as rigorous as several previous derivations [Dau2, Gro]. Those derivations employed limiting conditions to avoid functions such as the impulse function. This derivation assumes uses the impulse or delta function where, to be mathematically rigorous, it should use the, for example, a "sinc" function [Bos1] that has a bandwidth that approaches infinity in the limit (the limit makes an impulse function out of the sinc function). To avoid the limiting conditions and to significantly simplify the "derivation" for the intended audience, the author takes the liberty of utilizing an impulse function (again, more rigorous proofs are in the references).

<u>Continuous Inverse Wavelet Transform Theorem</u>

If $f(x)$ and $g(x)$ are finite energy functions (in $L^2(\mathbf{R})$) and $g(x)$ is an admissible mother wavelet with $g_{a,b}(x) = \frac{1}{\sqrt{|a|}}\, g\left(\frac{x-b}{a}\right)$ for $(a,b) \in (\mathbf{R}\setminus\{0\} \times \mathbf{R})$, then the **inverse wavelet transform,** W_g^{-1} , maps a surface in the scale-translation plane, $H(a,b)$, into the one-dimensional time or space domain ($W_g^{-1} : L^2(\mathbf{R}\setminus\{0\} \times \mathbf{R}) \rightarrow L^2(\mathbf{R})$) as:

$$\boldsymbol{W}_g^{-1} : H(a,b) \mapsto p(x)$$

$$p(x) = \frac{1}{C_g} \int_{-\infty}^{\infty} \int_{-\infty}^{\infty} H(a,b) \frac{1}{\sqrt{|a|}} g\left(\frac{x-b}{a}\right) \frac{db\,da}{a^2} \tag{2.7}$$

For the special case when the scale-translation surface is the wavelet transform of f *with respect to mother wavelet*, g, or $H(a,b) = \boldsymbol{W}_g f(a,b)$, then:

$$\boldsymbol{W}_g^{-1} : \boldsymbol{W}_g f\,(a,b) \mapsto f(x)$$

$$f(x) = \frac{1}{C_g} \int_{-\infty}^{\infty} \int_{-\infty}^{\infty} \boldsymbol{W}_g f\,(a,b) \frac{1}{\sqrt{|a|}} g\left(\frac{x-b}{a}\right) \frac{db\,da}{a^2} \tag{2.8}$$

The inverse transform creates the original function by summing appropriately weighted, scaled and translated versions of the mother wavelet. The weights are the wavelet coefficients, $\boldsymbol{W}_g f\,(a,b)$. Note that the inverse transform sums over the two-dimensional scale-translation space. The time or space domain function is created from the wavelet domain coefficients, $H(a,b)$ or $\boldsymbol{W}_g f\,(a,b)$, *and the mother wavelet*, g. Instead of just defining the inverse transform as equation (2.8), the more general form of equation (2.7) was employed to emphasize the dependence of the wavelet transform on its mother wavelet. If the wavelet transform was not taken with respect to the same mother wavelet, then the inverse transform with respect to a different mother wavelet will not necessarily reconstruct a function that even resembles the original time or space function. That is:

$$\frac{1}{C_g} \int\int \boldsymbol{W}_{g_2} f\,(a,b) \frac{1}{\sqrt{a}} g\left(\frac{x-b}{a}\right) \frac{da\,db}{a^2} \neq f(x)$$

$$\text{if } g_2(x) \neq g(x)$$

Returning to the functional approximation problem displayed in Figure 2.1, the inverse wavelet transform reconstructs the original signal from the wavelet coefficients as shown in Figure 2.7. Note that the reconstructing elements are the scaled and translated wavelets also, but they have not been complex conjugated (for the mathematically inclined, the **duals** of the wavelet transform elements are just the complex conjugates of these elements; however, this is true only for the continuous transform).

Next, the inverse wavelet transform is derived. The resolution of identity is applied to derive the continuous inverse wavelet transform for the wavelet transform defined in equation (2.2). Many of the derivations of the continuous inverse wavelet transform exploit the group theoretic structure to produce simple proofs [Gro2, Hei]. Although the power of the group theoretic mathematics simplifies wavelet theory, this book attempts to prove the required results without the requirement to know group theoretic mathematics. The derivation presented here simply exploits the delta function's sifting property.

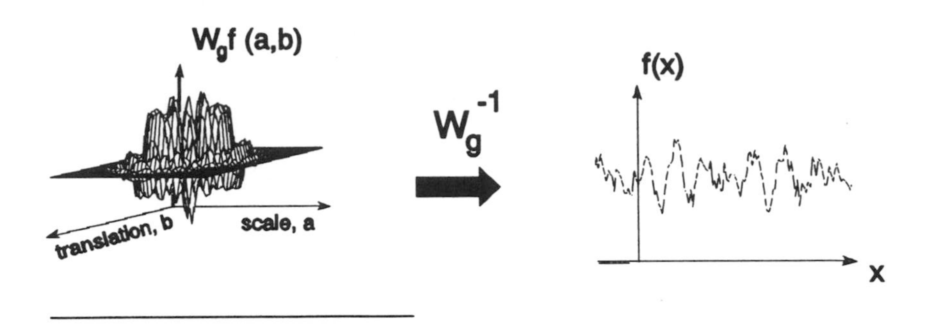

Figure 2.6: Inverse Wavelet Transform Example

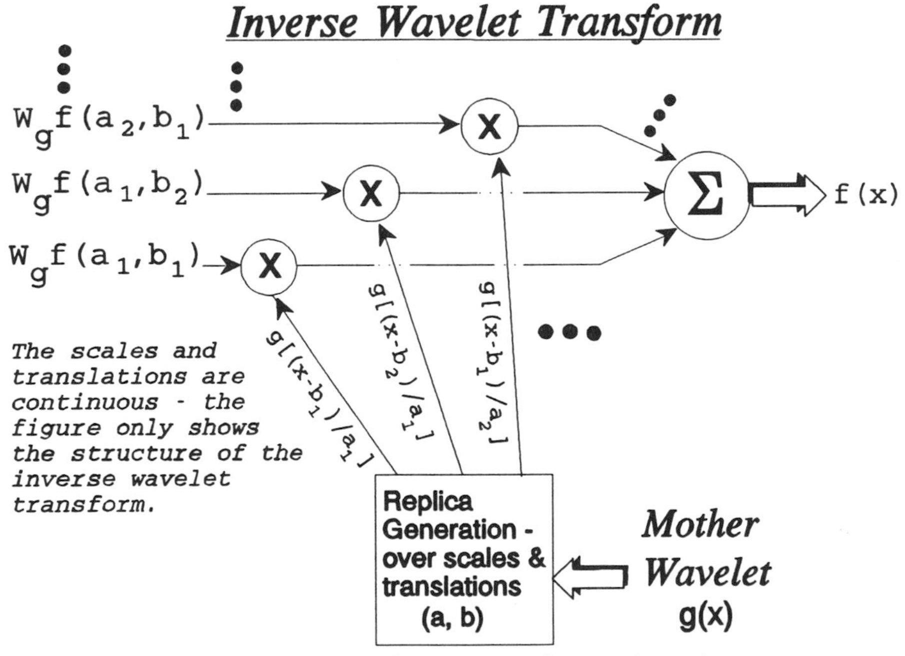

Figure 2.7: Inverse Wavelet Transform Structure

Continuous Inverse Wavelet Transform Proof: This proof applies the resolution of identity theorem but does not use any additional mathematics; for this reason the proof is not completely rigorous and more rigorous proofs can be found in [Dau, Gro2, Hei]. In the rigorous proofs the delta function is simply defined as a limit to avoid any subtle mathematical difficulties. Starting from the **resolution of identity** and substituting a translated delta function, $\delta(x'-x)$, for $f_2(x')$ yields:

$$\langle f_1, f_2\rangle = \langle f_1(x'), \delta(x'-x)\rangle = \int_{-\infty}^{\infty} f_1(x')\,\overline{\delta(x'-x)}\,dx' = f_1(x) \tag{2.10}$$

by the sifting property of the impulse function. Substituting this same impulse into the right hand side of the resolution of identity yields:

$$c_g\langle f_1, f_2\rangle = c_g f_1(x) = \int_{-\infty}^{\infty} \frac{1}{a^2}\int_{-\infty}^{\infty} \langle f_1, g_{a,b}\rangle \langle g_{a,b}, f_2\rangle\, db\, da \tag{2.11}$$

$$= \int_{-\infty}^{\infty} \frac{1}{a^2}\int_{-\infty}^{\infty} \langle f_1(x'), g_{a,b}(x')\rangle \langle g_{a,b}(x'), \delta(x'-x)\rangle\, db\, da$$

$$= \int_{-\infty}^{\infty} \frac{1}{a^2}\int_{-\infty}^{\infty} \langle f_1(x'), g_{a,b}(x')\rangle\, g_{a,b}(x)\, db\, da \tag{2.12}$$

$$= \int_{-\infty}^{\infty} \frac{1}{a^2}\int_{-\infty}^{\infty} \langle f_1, g_{a,b}\rangle \frac{1}{\sqrt{|a|}}\, g\!\left(\frac{x-b}{a}\right) db\, da \tag{2.13}$$

$$= \int_{-\infty}^{\infty} \frac{1}{a^2}\int_{-\infty}^{\infty} \mathbf{W}_g f_1(a,b) \frac{1}{\sqrt{|a|}}\, g\!\left(\frac{x-b}{a}\right) db\, da \tag{2.14}$$

therefore:

$$f_1(x) = \frac{1}{c_g}\int_{-\infty}^{\infty} \frac{1}{a^2}\int_{-\infty}^{\infty} \mathbf{W}_g f_1(a,b) \frac{1}{\sqrt{|a|}}\, g\!\left(\frac{x-b}{a}\right) db\, da \tag{2.15}$$

thus proving the (continuous) inverse wavelet transform.

The continuous wavelet transform and its inverse have now been stated and justified. Note that the mother wavelet is not conjugated for the inverse wavelet transform but it is conjugated for the wavelet transform in equation (2.2) (this is the dual relationship for the continuous transform).

Nonuniqueness of the Inverse Wavelet Transform

A characteristic of the inverse wavelet transform is that it is *nonunique* in the sense that *several, different, wavelet transform domain representations can be inverted with respect to the same mother wavelet to create the same time (or space) function.* The increase in the dimensionality of the representation causes this nonuniqueness. Due to the nonuniqueness, the coefficients at one set of scales and translations may be able to represent the coefficients at a different set of scales and translations. This states that the inverse wavelet transform may be a many-to-one operator; multiple wavelet domain representations map to the same time function. See Figure 2.22 for an example that is subsequently detailed in Appendix 2-A. Further consideration of this nonuniqueness is deferred to Appendix 2-A to avoid a discontinuity in the presentation of the wavelet theory. Later in the book, this nonuniqueness property is exploited to explain the properties of the wideband matched filter.

Energy Distribution in the Wavelet Transform Domain

Since the wavelet transform is an alternative representation of a signal, it should include all of the characteristics of the original time or space domain representation. One feature of the signal is its energy. In the time domain, the energy of a signal is the integral of its square over time:

$$E_f = \int_{-\infty}^{\infty} |f(t)|^2 \, dt$$

For a different signal representation, the Fourier representation [Van1], the energy of the signal can be characterized in the Fourier domain. Parseval's theorem [Van1] provides the energy relationship for the Fourier case. The differential amount of energy in a differential band of frequencies is the magnitude squared of the transform times the differential frequency band, $\Delta E_f = |F(\omega)|^2 \Delta\omega$ or $dE_f = |F(\omega)|^2 d\omega$. This energy distribution across frequency is often termed the spectrogram and is often used to estimate the power spectrum of a signal [Bou, Hla, Pap1, Rio]. When this differential is integrated over all frequencies, the energy of the signal results:

$$E_f = \frac{1}{2\pi} \int_{-\infty}^{\infty} |F(\omega)|^2 d\omega$$

For the wavelet transform an analogous differential energy can be defined. However, the differential "band" or area of the scale-translation plane is not just $da\,db$. The wavelet transform or scale-translation differential element is required to cause the energy to be the same in both domains. This differential element is $\frac{da\,db}{a^2}$ (derived subsequently). The energy distribution in the wavelet domain is termed the *scalogram* [Com, Rio] and is just the magnitude squared of the wavelet transform with the proper differential term. The total energy in the wavelet representation is:

$$E_f = \frac{1}{C_g} \int_{-\infty}^{\infty}\int_{-\infty}^{\infty} |W_g f(a,b)|^2 \frac{da\,db}{a^2} \qquad (2.16)$$

Thus, the differential energy in a differential area of the scale-translation plane or wavelet transform domain is $|W_g f(a,b)|^2 \frac{da\,db}{a^2}$ (the scalogram). Note that equation (2.16) has the inverse of the admissibility constant on the right hand side. The admissibility constant normalizes the energy of the scalogram. If the mother wavelet is normalized such that its admissibility constant is unity, then it can be ignored; otherwise, the constant must be included as justified by the following derivation of the energy distribution in the wavelet domain. In the spectrogram the kernel of the transform is the exponential function that has a magnitude squared of unity and can be neglected; in the scalogram or wavelet domain the energy distribution depends upon the mother wavelet. Besides the subtle normalization due to the mother wavelet, a much more significant dependence on the mother wavelet must be emphasized. *The scalogram will be different for each different mother wavelet - the scalogram is not a unique energy distribution for a particular signal; each different mother wavelet will distribute the energy differently across scale and translation plane.* Later, in Chapter 5, when systems are studied it will be demonstrated that the system's energy distribution is unique for each system. But, returning to a signal's energy distribution, the following justification is provided.

Justification of the Energy Distribution in the Wavelet Domain:

From the resolution of identity (equation (2.3)) with $f_1 = f_2 = f$:

$$c_g \langle f, f \rangle = c_g \int_{-\infty}^{\infty} |f(t)|^2 \, dt = c_g E_f$$

$$= \int_{-\infty}^{\infty}\int_{-\infty}^{\infty} \langle f, g_{a,b} \rangle \langle g_{a,b}, f \rangle \frac{db\,da}{a^2}$$

Since the inner products are just wavelet transforms:

$$E_f = \frac{1}{c_g} \int_{-\infty}^{\infty}\int_{-\infty}^{\infty} |W_g f(a,b)|^2 \frac{da\,db}{a^2} \tag{2.17}$$

justifying the wavelet domain energy distribution.

Other formulations of the wavelet transform (besides that in equation (2.2)) [Chai, Gro, Hei] can provide different energy distributions, but the most accepted wavelet transform definition is provided in equation (2.2) and the energy distribution defined here is consistent with that definition.

Discrete Wavelet Transform (Continuous Time Wavelet Series)

Analogous to Fourier theory, the continuous wavelet transform is not employed as often as the discrete wavelet transform. The continuous transforms are primarily employed to derive properties and the discrete forms are necessary for most computer implementations. The "discrete" wavelet transform is analogous to the Fourier series, not the discrete Fourier transform. The "discrete" term applies only to the transform domain parameter (i.e., the frequency variable for Fourier series and the scale and translation variables for the wavelet transform) and not the independent variable of the function that is being transformed (i.e. time, space, etc.). Alternatively stated, the scales and translations are discrete, but the independent variable is continuous (i.e., continuous space and/or time). Later, the independent variable will be discretized for computer implementations and the integration will be approximated by summations. For the aforementioned reasons, this transform should be defined as the *continuous time wavelet series* (CTWS).

The previous sections studied the continuous wavelet transform (continuous in the independent variable as well as scale and translation). This section describes the discrete (in scale and translation) wavelet transform. A discrete wavelet transform (or CTWS) yields a countable set of coefficients in the transform domain. The coefficients correspond to points on a two-dimensional grid or lattice of discrete points in the scale-translation domain. This lattice will be indexed by two integers, the first integer, m, will correspond to the discrete scale steps while the second integer, n, will correspond to the discrete translation steps (the grid is indexed by m and n). The scale, a, is now $a=a_o^m$ and the translation, b, is $b=nb_oa_o^m$, where a_o and b_o are the discrete scale and translation step sizes, respectively.

The discrete wavelet transform is defined with respect to a continuous mother wavelet, g. The discrete wavelet transform maps $\mathbf{W}_g: L^2(\mathbf{R}) \rightarrow l^2(\mathbf{Z}^2)$ (continuous finite energy signals to a 2-D discrete grid of coefficients - Z represents the set of all positive, negative and zero integers). The **discrete wavelet transform or continuous time wavelet series (CTWS)** is:

$$\mathbf{W}_g f(m,n) = \frac{1}{\sqrt{a_o^m}} \int_{-\infty}^{\infty} f(x)\, g\left(\frac{x-nb_oa_o^m}{a_o^m}\right) dx = a_o^{\frac{-m}{2}} \int_{-\infty}^{\infty} f(x)\, g(a_o^{-m}x-nb_o)\, dx$$

$$= \langle f, g_{m,n}\rangle = \langle f, U(a_o^m, nb_oa_o^m)\, g\rangle \tag{2.18}$$

where the unitary affine operator, U, was previously defined in the continuous wavelet transform. Note that the processing is over continuous time but the wavelet representation is on a discrete grid. These discrete wavelet coefficients represent the original signal but, as in the continuous case, the representation is sensitive to the chosen mother wavelet. Unlike the continuous wavelet transform, the discrete wavelet transform is defined only for positive scale values, $a_o > 0$. This constraint on the scale is not restrictive because the reflected mother wavelet (a scale of -1) can be used as the new mother wavelet and effectively cover negative scales as well.

The corresponding continuous translation steps are $b=nb_oa_o^m$ so that the translation steps are proportional to the time scaling (see Figure 2.8). Refer back to Figure 1.4-Figure 1.5; for large scale values the wavelet is a dilated version of the mother wavelet (the time support is large, thus leading to poor time resolution - this corresponds to resolution cells at the bottom of Figure 2.8). For small scale values the mother wavelet is compressed (leading to small time support and good time resolution, but poor frequency resolution - corresponding to the resolution cells at the top of the figure). Figure 2.8 is another description of the time-frequency or time-scale resolution characteristics of these scaled and translated mother wavelets. For this figure the *scale is increasing in the downward direction* while the frequency increases in the upward direction (the inverse relationship between scale and frequency was demonstrated in Chapter 1).

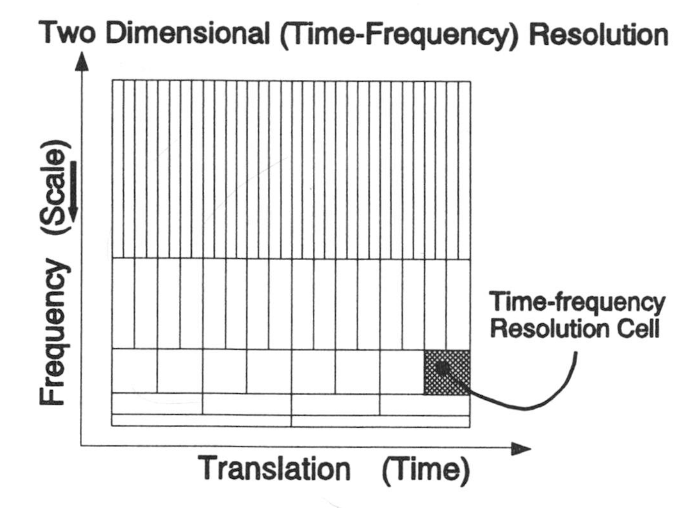

Figure 2.8: Time-scale Resolution of Wavelet Transforms

Only the wavelet transform resolution *with respect to the mother wavelet in Figure 1.1* is depicted in Figure 2.8; the purpose of the figure is to demonstrate a particular scale-translation lattice for the CTWS. Figure 2.8 has several interpretations. The size of each rectangle in this figure is dictated by the size of the particular scaled and translated mother wavelet (as discussed in the previous paragraph). Each rectangle represents the simultaneous time-scale *resolution* of each scaled and translated mother wavelet function; note that the resolution cell size changes for each scale but is identical for a fixed scale and different translations; this is analogous to "constant-Q" filtering in which the fractional bandwidth, the ratio of bandwidth to center frequency, is held constant across multiple octave changes in the center frequency. Going "up" the figure would correspond to higher center frequencies and the width of the resolution cell in frequency (bandwidth) increases maintaining the constant ratio (Q) between the center frequency and the bandwidth. If this figure is viewed "from the left side," then the constant fractional bandwidth division of the frequency axis becomes clear - lower frequency bands keep being divided in half (Figure 2.15 demonstrates this constant-Q division as well).

Another feature of Figure 2.8 is that it represents how the wavelet coefficients are generated and ordered. As discussed in the last paragraph, each resolution cell represents a different scaled and translated mother wavelet. Since each different scale and translation leads to a different wavelet coefficient (see Figure 2.7), each resolution cell also represents a discrete coefficient in the scale-translation space (a wavelet domain coefficient, $W_g(m,n)$). The wavelet coefficients corresponding to large scales require longer time durations to compute due to the dilated mother wavelets stretching out. Thus, the coefficients corresponding to larger scales are not output as often as those corresponding to smaller scales which require less time to compute. Therefore, if Figure 2.8 is traversed along the time axis, then the wider resolution cells in time will have a coefficient generated less often and the figure dictates exactly how often. Thus, by traversing the figure in the direction of advancing time the figure indicates the times (and the corresponding scale) at which a new coefficient will be output from the transform. This "generating a coefficient less often" leads to the *subsampling* (throwing away intermediate coefficients) at the larger scales or lower frequencies. The small scale or higher frequency coefficients are computed much more often than the larger scale wavelet coefficients (subsampling is detailed further later in this chapter).

For a continuous wavelet transform, with respect to an admissible mother wavelet, the finite energy function, f(x), always has a convergent reconstruction from its set of continuous wavelet coefficients $\{\mathbf{W}_g f(a,b) : (a,b) \in (\mathbb{R}\backslash\{0\},\mathbb{R})\}$. When the wavelet coefficients are evaluated only on a two-dimensional lattice of points (CTWS), then the reconstruction may be unstable (not convergent) - it depends on the choice of the mother wavelet, g, *and the density of the discrete lattice*. The density of the lattice is determined by the **lattice step sizes** $\{a_o, b_o a_o^m\}$ which define a lattice structure as in Figure 2.8. Note the logarithmic sample structure of the lattice in both dimensions. This is due to the exponentially changing scale, a_o^m. The lattice step sizes are very important and will dictate many of the properties of discrete wavelet transforms including the form of its inverse and the set of allowable mother wavelets.

Wideband Matched Filter Interpretation of the Lattice Density

Since matched filter theory [Van1] is more broadly understood than wavelet theory, an analogy between matched filter processing and wavelet transforms may help motivate and focus the mathematics that follow in the next several sections. Matched filtering can generally be stated as the design of an optimum filter for a received signal. One of the primary considerations in the implementation of a matched filter is the density of the hypothesized parameters. For efficient implementations it is desirable to have a sparse set of hypotheses, but to avoid any losses in gain and/or resolution, the density of hypotheses should be tight. Thus, tradeoffs exist for the set of hypothesized parameters or denseness of the hypothesis "grid."

Before proceeding with the detailed, rigorous, and highly mathematical justification of the appropriate sampling density (lattice) for the discrete wavelet transform, an intuitive interpretation of these conditions is stated with a filtering system example. Some researchers [Dau, Hei] have considered how sparse this discrete lattice can become. For some practical situations the sparseness of the lattice is limited by the characteristics of the signals and/or systems and their sensitivity to noise or interference. Very sparse lattices ($a_o = 2$) can be inappropriate for some signals and/or systems; in many practical situations a dense lattice will be chosen because noise will interfere as the lattice becomes sparse.

The sparseness of the scale-translation points in the wavelet domain can be related to the sparseness of hypotheses in matched filter theory, where the question is: How close do the delay and Doppler hypotheses (sampling density of the delay-Doppler plane) have to be? (This question is answered with ambiguity function analysis and the associated signal resolution [Van2]). The delay and Doppler hypotheses are analogous to the scale-translation values. Each resolution cell (coefficient) can be considered as the output of a matched filter that uses the particular scale and translation to create a "replica" or filtering signal (the scaled and translated mother wavelet determines the filter impulse response). See Figure 2.9.

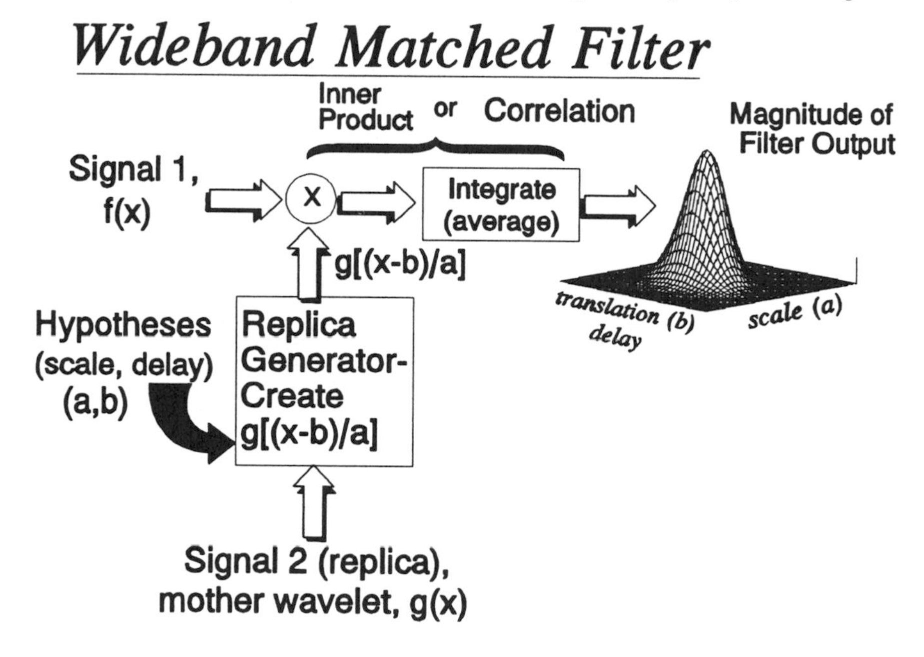

Figure 2.9: Wideband Matched Filter

The sparseness of the scale and translation sample points sets a limit on the "acceptable loss" in signal energy; how much energy can be missed and still lead to an acceptable representation? Figure 2.10 presents an example of the matched filter output for a particular input signal (the input signal is a mother wavelet for this case). Each intersection of the lines in Figure 2.10 represents a sample point in the scale-translation plane. Judging from the smoothness of Figure 2.10, the sampling was dense enough to provide an acceptable representation (intermediate points will not be significantly higher than any of the sample points). If, instead, the intersection of every tenth line represents a sample point (see the big black dots on Figure 2.10), then the representation would look almost like 4 impulses or a small plateau. The highest point on this more sparsely sampled grid could be well below the peak; almost 20 dB lower for the specific case that is shown. The less dense grid would most likely be an unacceptable wavelet representation for many applications.

Since practical processes involve noise, the noise will dominate any of the "low lying" energy in the scale-translation or wavelet domain plane. So, one limit on the sparseness is that the lattice should be dense enough to cover the entire plane at some selected level (3-10 dB down from the peak points). For the sparsely sampled wavelet representation (every tenth sample) discussed in the previous paragraph, the wavelet domain representation may be very sensitive to noise perturbations. This interpretation is intuitive, but for those familiar with matched filter theory, it should be acceptable. Thus, the sparseness of the lattice controls the gain and noise sensitivity of the matched filter output and will similarly control these properties for the wavelet domain representation.

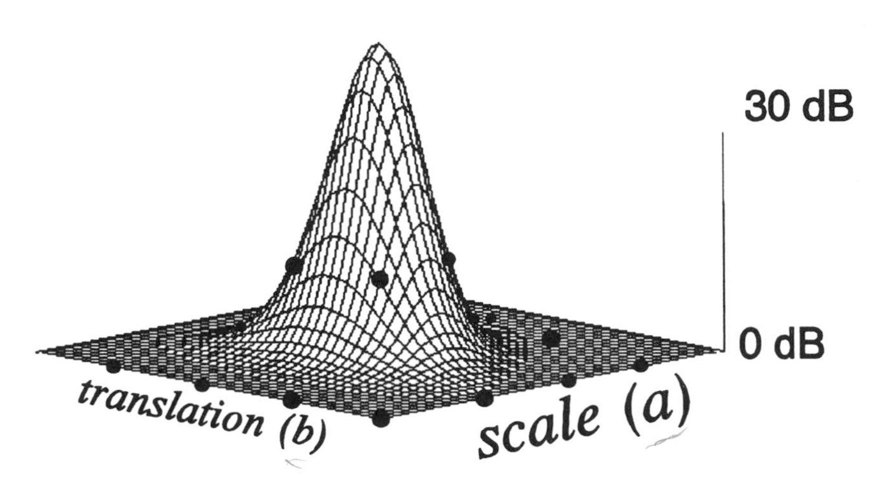

Figure 2.10: Magnitude of Matched Filter Output

Later in Chapter 3, active sensing will be considered and then the transmitted signal is used as the mother wavelet in the wideband matched filter (see Figure 3.9); thus, Figure 2.10 would correspond to the output of a wideband matched filter when the reflected signal is an exact replica of the transmitted signal. This matched filter output corresponds to one resolution cell in Figure 2.8; if the surface in Figure 2.10 is sliced by a plane that is 3 dB down from the peak, then an ellipse will be formed at the intersection of this plane and the surface. One of the rectangles in Figure 2.8 approximates this ellipse. The other rectangles are approximated by using scaled and translated mother wavelets as the "signal 1" or f(x) in Figure 2.9 and slicing each of the corresponding output surfaces at their 3 dB down points; each different scale and translation leads to a different rectangle. It is important to understand that the size of these rectangles will change when the surface is sliced at some other (N dB down) point. If the grid remains the same but these surfaces are sliced at the 30 dB down points, then each rectangle will be much larger and they will overlap. The overlap suggests redundant information (at the 30 dB down level) for this particular grid. If 30 dB of loss is acceptable, then the grid can be sampled much more sparsely (requiring less processing). The new sampling density can be set by moving adjacent grid sample points such that the 30 dB rectangles or ellipses just overlap. The set of overlapped output surfaces forms a composite "lumpy" surface across the scale-translation plane and it is the intersections of each of the individual output surfaces that form the lines dividing the resolution cells in Figure 2.8. Thus, the sampling grid density is set by the resolution of the signal and the acceptable loss or distortion.

If an input, f(x), is a scaled and translated version of the mother wavelet but does not correspond to one of the exact scale-translation hypotheses (the center of the resolution cells), then it will fall somewhere on this lumpy surface and will *never be more than 3 or N dB below the peak(s) of the composite surface*. The signal gain at any point on the surface will not be lower than some minimum value (arbitrarily selected as 3 or N dB down from the maximum gain). If the grid is selected to have intersections that are 30 dB down from the peak instead of just 3 dB, then the hypotheses can be much more sparse. However, the more sparse the lattice of hypotheses becomes, the higher the possible loss becomes and the more sensitive the estimator becomes to noise. Thus, the sparseness of the lattice controls the gain and noise sensitivity of the matched filter output and will similarly control these properties for the wavelet domain representation.

An important feature of these resolution surfaces, as in Figure 2.8, is that both the translation and scale resolution change as the scale hypothesis changes. When small scales are hypothesized, then the matched filter (and wavelet transform) have good translation resolution and poor scale resolution relative to large scale hypotheses. Later it will be demonstrated that the resolution cell for the *delay-Doppler (narrowband)* matched filter's output does not change shape (not scaled) - it will be shown to be invariant across Doppler shifts (see Figure 3.2). The wideband matched filter's resolution properties change (are not invariant) as different scale values are hypothesized. It will also be demonstrated that Doppler shifting only approximates time scaling. Further comparison and analogies between wavelet theory, matched filter theory, and ambiguity functions will be detailed in Chapter 3. Next the effect of these discrete lattices on the mother wavelet is considered.

Review of the Inverse Discrete Wavelet Transform: Scale-translation Lattice Density and Mother Wavelet Constraints

The discrete wavelet transform or continuous time wavelet series (CTWS) in equation (2.18) is analogous to the bank of matched filters discussed in the previous section. A new filter is formed for each distinct scale and translation combination. The discrete wavelet transform represents a continuous function of time by a countable set of wavelet coefficients, $W_g(m,n)$. The wavelet coefficients only exist at discrete points in the scale-translation plane (discrete grid). An integer index for both scale and translation define those points. The discrete scale and translation values are: scale: $a = a_o^m$, and translation: $b = nb_o a_o^m$, so that the discrete indices are m and n for scale and translation, respectively. For the discrete wavelet transform the integer indices, m and n, replace the continuous scale and translations variables, a and b, respectively.

The scale and translation discrete step sizes, a_o and b_o, respectively, determine the density of the discrete lattice. The translation discretization step size, b_o, controls how far the mother wavelet is slid or translated before the next wavelet coefficient is computed. If the translation step size, b_o, becomes too big, then the translated mother wavelet that is one translation step away can look very different than a non-translated mother wavelet; the correlation between the function f and these two wavelets can be very different (the wavelet domain surface may no longer be smooth). *Thus, the choice of* b_o *will depend upon the translation or time resolution properties of the chosen mother wavelet (later this resolution is described with ambiguity functions). An analogous situation holds for the scale discretization step size,* a_o*, which should depend upon the scale resolution of the mother wavelet.*

By considering the mother wavelet as a lowpass or bandpass signal, a minimum sample rate can be established for this signal. For some applications this sampling rate (appropriate for the mother wavelet) can be used as the b_o. Then b_o will act as the sample period (in space or time). Thus, b_o can be specified by sampling criteria such as the Nyquist criteria; the mother wavelet with a scale of unity is the signal that must be sampled sufficiently fast. Since the mother wavelet can have multiple frequency components, simultaneously or in succession, a general sampling rule for all mother wavelets cannot be stated. Other considerations or priori knowledge regarding signal characteristics can also affect the choice of b_o. A more detailed discussion of the resolution properties of the mother wavelet is deferred until Chapter 3 when the ambiguity functions are discussed.

Often, the scale discretization step size, a_o, is chosen to be 2 so that discrete time signals can be easily scaled - for a sampled signal, scaling by a factor of 2 is implemented by taking every other sample point to form the scaled signal (decimating or subsampling). Although the scaling factor of 2 (octave band filters) is desirable for representing some signals it may not be desirable for representing other signals or for representing system actions. Again, details are provided later. A technique termed **voicing** is an intermediate step. Voicing uses multiple wavelet transforms that act together to effectively form one wavelet transform at scale steps that are finer than powers of 2. The multiple transforms each have a different mother wavelet; however, each different mother wavelet is just a scaled version of a single admissible function (Grossman, Kronland-Martintet, and Morlet's article in [Com]). When the multiple wavelet transforms are considered together, an effectively higher scale resolution is achieved (but the resolution in translation has remained the same). Voicing can provide better resolution and a consistent analysis across finer scales while maintaining the efficient factor of 2 scaling. Voicing is later compared to other multiple mother wavelet transforms.

Once a discrete scale-translation lattice is established then a mother wavelet is chosen. Desirable properties for the mother wavelet constrains the set of allowable mother wavelets. In the extreme case when the constraints are: $a_o = 2$, orthogonal wavelets on that lattice, smooth mother wavelets, and compactly supported mother wavelets (nonzero only for a finite length of time) then the wavelets become unique [Dau4], and only one mother wavelet exists satisfying those conditions. As illustrated by this example, the constraints can be extensive. For the applications to be discussed later, it will be desirable to allow the set of mother wavelets to be as large as possible (for multiple mother wavelet representations and for a system representation in which the input signal becomes formulated as a mother wavelet) and, thus, a minimum of constraints is desirable.

For applications with $a_o \approx 1$ the scale-translation lattice becomes dense and the resolution characteristics of the scaled and translated mother wavelets will overlap (become redundant). The overlapping regions are represented by several different scaled and translated mother wavelets. Even when $a_o \approx 1$ the scales can still cover multiple octaves (i.e., if $a_o = 1.1$ then just 50 values of m will cover 7 octaves) and, thus, the condition that $a_o \approx 1$ does not limit this lattice to only represent narrowband signals (7 octaves signals are definitely wideband). For those who read the next section (that is the most mathematical section in the book), the $a_o \approx 1$ condition is restated by requiring *"snug* frames." After ambiguity functions are presented and related to the resolution of the wavelet transform it will become obvious that the correct choice for the lattice density (a_o and b_o) depends upon the resolution properties of the mother wavelet.

Under the condition $a_o \approx 1$ and b_o is chosen such that it samples the mother wavelet at least at the Nyquist rate (these assumptions provide a dense lattice in the wavelet domain), the appropriate **inverse discrete wavelet transform** becomes:

$$f(x) \approx k \sum_{m=0}^{\infty} \sum_{n=0}^{\infty} [\boldsymbol{W}_g f(m,n)]\, a_o^{\frac{-m}{2}}\, g(a_o^{-m}x - nb_o) \qquad \textbf{(2.19)}$$

where k is a constant that depends upon the redundancy of the mother wavelet and the lattice combination (k will be greater than one if the lattice and mother wavelet combination are highly redundant (the resolution cells overlap at the 3 dB down points) and k will be approximately one as the overlap decreases (30 dB down overlap)). This constant is normally ignored in many applications as it does not affect the relative weighting of the coefficients - it simply multiplies the reconstructed signal. The value of k can be determined and its calculation is deferred to the next section, $k = \frac{2}{A+B}$ where A and B are the frame bounds.

Note that equation (2.19) is an approximation. This approximation is good as long as the wavelet domain representation maintains some redundancy (some overlap of the resolution cells). If the lattice is made more sparse (a larger scale step size) then the wavelet domain representation becomes less redundant (little overlap of the resolution cells) and the reconstruction of equation (2.19) can become invalid; the mother wavelet becomes the wrong function to perform this reconstruction (the "dual" of the mother wavelet, as discussed in the next section, is required). This condition is avoided in this book. For all of the processing considered in this book, the lattice will be designed to be dense enough to make the approximation in (2.19) valid.

The inverse discrete wavelet transform has a structure that is identical to the inverse *continuous* wavelet transform and, thus, the discussion of the reconstruction process is not repeated here.

A discussion of the choice of the mother wavelet and their impact on the discrete wavelet transform is deferred until a discussion of the resolution properties of mother wavelets and their interpretation as ambiguity functions. Next, the mathematically rigorous discussion of the discrete wavelet transform is addressed.

Discrete Wavelet Mathematics - Rigorous Justification

This section does not follow the requirement imposed on the rest of the book; some specialized mathematics are required to follow the analysis done in this section. That is the justification for summarizing the results of this section in the previous section. This section is included to provide a more complete treatment of wavelet theory for those that have the necessary mathematical background.

The detailed mathematics in this section are necessary to derive the requirements for choosing mother wavelets and to determine the "sparse" scale-translation lattices appropriate for discrete wavelet transforms. Further details and more rigorous derivations can be found in the references [Chu, Dau, Hei]. The primary result of this section is that the lattices cannot be made arbitrarily sparse without considering the mother wavelet properties. However, for those who prefer to initially avoid the more detailed mathematics in this section, a less mathematical (and less thorough) review section was provided in the previous section. The previous section summarized the essential concepts of this section. The summary is adequate to follow the discussions later in the book. If desired, skip this section on the first reading.

This section follows the research of Daubechies [Dau] very closely. For a given lattice structure $\{a_o, b_o\}$ the possible mother wavelets that are admissible and lead to a stable reconstruction are limited. The conditions which yield acceptable mother wavelets are derived in terms of **frames**. A *frame* [Duf, Hei] is a set of functional elements with the special property that any non-zero function must have a non-zero projection onto at least one of these elements (a complete set) and any non-infinite function must have a sum of projections that is less than infinity. However, a common property of frames that distinguishes them from common vector spaces it that the elements can be linearly dependent or redundant. It will be required that the discretely scaled and translated mother wavelets, $\{g_{m,n}; m, n \in \mathbf{Z}\}$, constitute a frame. However, even if a mother wavelet leads to a frame, the inverse discrete wavelet transform may still be unacceptable due to the formation of the reconstructing elements (dual elements). For some mother wavelet and lattice combinations the reconstructing elements are not simply scaled and translated versions of the mother wavelet function - the reconstructing elements can change as the translation changes. These mother wavelet and lattice combinations are undesirable. Orthogonal, biorthogonal and related wavelet transforms have addressed the problem of assuring that the reconstructing elements are scaled and translated versions of the mother wavelet. For the applications discussed later, these "orthogonal" transforms may not be acceptable and, thus, an alternative condition requiring the scaled and translated mother wavelets as the reconstructing elements is stated.

A *frame* is a set of elements in a Hilbert space which satisfy the frame bounds for all elements of the Hilbert space. A set of vectors $\{\phi_j; j\in J\}$ (for the wavelet frames the vectors will be the scaled and translated mother wavelets, $\phi_j = g_{m,n}$, over different index sets) in a Hilbert space, **H**, for which the sum $\sum_{j\in J} |\langle f, \phi_j\rangle|^2$ of a non-zero signal, $f\in\mathbf{H}$, has a finite upper bound and a nonzero lower bound is called a frame. These vectors, ϕ_j, are termed the frame elements (they are not basis vectors in general).

The frame bounds, A and B, are defined by:

$$A\|f\|^2 \le \sum_{m,n} |\langle f,\phi_j\rangle|^2 \le B\|f\|^2 \tag{2.20}$$

for every nonzero element, f, of the Hilbert space with $A>0$, $B<\infty$ and A,B both independent of f. This obviously requires that the set of elements be complete (no nonzero element of the Hilbert space has all zero coefficients) due to the lower bound. Stability of the reconstruction is considered as a required condition - this is where the "new" wavelet interpretation is different from the previous Gabor-type of group transforms.

The distinction between these frames and usual, orthogonal bases is that a frame does not necessarily consist of linearly independent elements. The dependence between elements creates a redundancy in the representation. This suggests that the representation need not be unique - but the representation may be constrained to be unique in a different sense by allowing only the "minimal" solution [Dau4]. This is further developed after introducing the operator formulation for this frame structure.

The frame example shown by Daubechies [Dau2] best illustrates the properties of a frame. Consider $\mathbf{R}^2$ and its elementary orthonormal basis, e_1 = [1 0] and e_2 = [0 1]. Define 3 new normalized elements in $\mathbf{R}^2$:

$$\phi_1 = e_1, \quad \phi_2 = \frac{-1}{2}e_1 + \frac{\sqrt{3}}{2}e_2, \quad \phi_3 = \frac{-1}{2}e_1 - \frac{\sqrt{3}}{2}e_2$$

Due to their orientation and normalization, every vector, f, in $\mathbf{R}^2$ will satisfy:

$$\sum_{j=1}^{3} |\langle f,\phi_j\rangle|^2 = \frac{3}{2}||f||^2 \quad and \quad f = \frac{2}{3}\sum \langle f,\phi_j\rangle\, \phi_j$$

which demonstrates the frame bound of 3/2 (A=B=3/2) and the redundancy that can exist in a frame.

Under special conditions, frames may be orthonormal bases. Daubechies, Meyer, Mallat, Battle and others all construct orthonormal wavelet bases. Frames are required for a wavelet transform but other constraints may be imposed as well. The purpose of the frame was to provide the existence and structure of the reconstruction; after a mother wavelet is determined, that wavelet and its scale-translation lattice points create a frame. A discrete wavelet transform does not require any further constraints. Determining the mother wavelets which create a frame may be easier if more constraints are introduced. Researchers have concentrated on finding mother wavelets which create some type of constrained frame [Dau, Hei, Mal, Mey, Vet]. As mentioned above, one desirable constraint is orthonormality of the wavelet transform frame elements. This orthonormality may be further constrained to provide additional structure to the frame elements. Compactly supported mother wavelets and regularity (which controls the rate of convergence) conditions are examples of additional constraints. The constraints on the mother wavelets are detailed further in later sections of this chapter and Chapter 3, but first the frame operator and its properties are presented.

Wavelet Transform Frame Operator

This section introduces the frame operator [Dau] and derives the discrete wavelet transform reconstruction formula. In a later section this operator point of view is used to simplify the analysis for demonstrating the noise or sensitivity reduction properties of wavelets. The dual elements which were previously introduced are discussed in detail. The dual elements for the continuous wavelet transforms are just the elements themselves. Frame operators are not required in the continuous case.

Using the previously presented frame concepts, define the frame operator, T, associated with the frame $(\phi_j)_{j\in J}$ as a linear operator which maps $T: H \to l^2(J)$ by $(Tf)_j = \langle f, \phi_j \rangle$. Here $l^2(J)$ stands for the space of square summable complex sequences indexed by J and J is the lexicographically ordered set of scale and translation indices, m and n, respectively. Note that this definition of the frame operator follows the notation presented in [Dau4] and does not follow the notation presented in [Hei]. The operator T is clearly bounded, $\|Tf\| \le B^{\frac{1}{2}} \|f\|$. Its adjoint operator T^* maps $l^2(J)$ onto **H**; it is defined by $T^* c = \sum_{j\in J} c_j \phi_j,$ *where* $c = (c_j)_{j\in J} \in l^2(J)$. The frame inequalities can be rewritten as:

$$AI \le T^* T \le BI \tag{2.21}$$

where I is the identity operator (If=f for all f in the Hilbert space). Since the Hermitian operator T^*T is bounded from below by a strictly positive constant, it is invertible, with a bounded inverse. This inverse satisfies:

$$B^{-1} I \le (T^* T)^{-1} \le A^{-1} I \tag{2.22}$$

Define the **dual frame** elements of the frame, $\{\phi_j : j \in J\}$, as $\tilde{\phi}_j = (T^*T)^{-1} \phi_j$. Then the set, $\{\tilde{\phi}_j\}_{j\in J}$, constitutes another frame. When the frame bounds are nearly equal (tight frames) then the dual element is approximately a constant times the original element.

The primary properties of the frame operator are [Dau4]:

1) T is a continuous linear operator: if f and g are both functions in **H** which are close in the **H** norm, ($\|f-g\| \le \epsilon$), then the sequences $(Tf)_j$ and $(Tg)_j$ will also be close in the l^2 norm:

$$\|Tf-Tg\|^2=\sum_{j\in J}|\langle f,\phi_j\rangle-\langle g,\phi_j\rangle|^2 \le B\epsilon^2 \qquad (2.23)$$

2) T is one-to-one: $Tf=Tg$ implies $f=g$.
3) T has a continuous inverse: if the sequences Tf and Tg are close in the l^2 norm, then f and g must have been close in the **H** norm:

$$\|f-g\|^2 \le A^{-1}\|Tf-Tg\|^2 \qquad (2.24)$$

4) The set, $\{\tilde{\phi}_j\}_{j\in J}$, is a frame with bounds B^{-1} and A^{-1}.
The first three properties are equivalent to the requirements for a frame (using the definition of the frame operator as well). Property 3 establishes the existence of a numerically stable reconstruction algorithm of a function in **H** from the l^2 sequence of projections onto the frame elements. Property 3 is not always satisfied in Gabor transforms which explains their possible instability.

The frame operator and its adjoint operator are used to derive the discrete wavelet transform reconstruction formula (inverse discrete wavelet transform):

$$f=(T^*T)^{-1}(T^*T)f=(T^*T)^{-1}\sum_{j\in J}(Tf)_j\phi_j$$

$$=(T^*T)^{-1}\sum_{j\in J}\langle f,\phi_j\rangle\phi_j=\sum_{j\in J}\langle f,\phi_j\rangle(T^*T)^{-1}\phi_j$$

$$f=\sum_{j\in J}\langle f,\phi_j\rangle\tilde{\phi}_j \qquad (2.25)$$

where $\tilde{\phi}_j$ are the dual frame elements. During the derivation it was assumed that $(T^*T)^{-1}$ commuted with the sum over j; this is valid due to the definition of the frame operator and the first three properties discussed earlier in this section. By simply rearranging the order of the operators it can also be shown that:

$$f=\sum_{j\in J}\langle f,\tilde{\phi}_j\rangle\phi_j \qquad (2.26)$$

Note that the dual functions are the "reconstruction elements." *These dual functions are not necessarily scaled and translated versions of the same function* as in the discrete wavelet transform. This book concentrates on applications that require redundant frames (nonorthogonal wavelets) and a dense lattice which will allow the dual elements to be formed by scaling and translating a single function, and specifically, the mother wavelet used in the wavelet transform.

From these general frame concepts the discrete wavelet transform can be derived [Dau, Hei, Mal]. The complete derivation is in those references. This book is concerned with the properties of the mother wavelet and the constraints associated with their construction, and thus further mathematical details of the discrete wavelet transform derivation are not repeated here. In general, the frame elements are replaced by scaled and translated versions of the mother wavelet, and the reconstruction formula in equation (2.25) will be valid up to a scalar constant, the inverse of the lower frame bound, A^{-1}.

With the discrete wavelet transform reconstruction formula the dual elements must be computed. In the continuous wavelet transform the dual elements are simply the elements themselves and a new space is not required. The computation of the dual elements requires knowledge about the frame bounds. In general, as the frame becomes more redundant, or the lattice becomes more dense, the frame becomes tighter. For tight frames the dual elements are simply constant multiples of the original frame elements; thus, the inverse discrete wavelet transform elements will just be the same scaled and translated mother wavelets used in the wavelet transform and the entire inverse transform will be multiplied by a constant to account for the redundancy. Again, details are found in the references but this book concentrates on redundant, nonorthogonal wavelets that have relatively dense scale-translation lattices and thus nearly tight frames (termed snug frames by Daubechies). Chui's book [Chu] concentrates on the octave scales and provides an excellent presentation of the duality and orthogonality conditions.

Similar to the continuous wavelet transform the inverse discrete wavelet transform is not unique either. But its nonuniqueness acquires an added dimension for the discrete case; the coefficients in the transform domain (scale-translation) can be the same but different dual elements can lead to the same reconstructed signal. This condition must be resolved for the wavelet operator to have a unique inverse. It is resolved by constraining the reconstruction to be a unique "minimal" reconstruction. The minimal condition constrains the dual functions to have the minimal "energy" (L^2 norm) possible. These minimal solutions are obtained by forming the dual elements as $\tilde{\phi}_l = (T^*T)^{-1}\phi_l$ - nonminimal dual elements would add a constant to this element. Calculation of these dual elements is one of the several recursions in wavelet processing. The formation of these dual elements from the frame elements is [Dau4]:

$$\tilde{\phi}_l = \frac{2}{A+B}\sum_{k=0}^{\infty}\left(I-\frac{2(T^*T)}{A+B}\right)^k \phi_l \tag{2.27}$$

which is derived by defining $S = T^*T$ and establishing the identity:

$$S = \frac{A+B}{2}\left[I-(I-\frac{2S}{A+B})\right] \tag{2.28}$$

and using the infinite sum formula:

$$(I-Q)^{-1} = \sum_{k=0}^{\infty} Q^k; \ \|Q\|<1 \tag{2.29}$$

the summation can be rewritten and placed into the middle of the frame bound equation (2.20) to yield:

$$\frac{A-B}{A+B}I \le \left[I-\frac{2S}{A+B}\right] \le \frac{B-A}{A+B}I \tag{2.30}$$

or

$$\left\| I - \frac{2S}{A+B} \right\| \leq \frac{B/A-1}{B/A+1} < 1 \qquad (2.31)$$

Thus, the series (equation (2.27)) for the dual frame in terms of the original frame elements converges, and the closer B/A is to 1 (tightness) the faster the series will converge. The frame tightness is generally controlled by the lattice density for a given mother wavelet and by choosing a reasonably dense lattice the dual elements can be quickly approximated. Then these dual elements can used to compute the inverse discrete wavelet transform. When the frame is snug, then only the first term in the series expansion of the dual frame elements (equation (2.27)) will be significant and, thus, the dual element in equation (2.26) can be replaced:

$$\tilde{\phi}_j \approx \frac{2}{A+B}\phi_j \quad and \quad f \approx \sum_{j \in J} \langle f, \phi_j \rangle \phi_j \qquad (2.32)$$

Several of the constraints imposed on mother wavelets and the scale-translation lattices are formulated in terms of frame concepts. *Tight frames* are defined as a frame which has equal frame bounds, A=B as in the example presented earlier. *A frame is termed* ***snug*** *if $A \approx B$ (this is the primary frame requirement enforced in this book).* For tight frames the reconstruction becomes $f=A^{-1}T^{*}(Tf)$ where the (Tf)'s are the wavelet coefficients and the inverse transform is $f=A^{-1}\sum_{m,n} \langle f, g_{m,n} \rangle g_{m,n}$ This sum converges strongly to f as the number of coefficients approach infinity [Dau1]. Note that frames may not be bases of a vector space. They usually have "extra" vectors which create a linearly dependent set of vectors. A frame is an *exact frame* if it ceases to be a frame whenever any single element is deleted from the frame. These frames are not redundant and are de-emphasized in this book. The redundant frames will tend to be snug and will be less sensitive to noise. The practical applications and the impact of redundant frames is discussed in the applications.

Now if a frame is tight, exact and A=B=1, then the frame constitutes an orthonormal basis. A possible advantage of redundant frames over orthonormal bases is that the less constraining requirements allow wider class of functions to be considered as mother wavelets. Other computational noise gains of redundant frame representations have been established [Daub4]. Later, in Chapter 4, the book derives efficient operators that suggest utilizing multiple mother wavelets. It is desirable to keep the set of allowable mother wavelets as large as possible. The redundant frame vectors may better match the signal content and only a small number of coefficients may represent a majority of the signal's information content. Thus, instead of requiring orthogonality and imposing constraints on the mother wavelets, the scaling step size is chosen so as to create a dense lattice and maintain the freedom of choice for the mother wavelet.

Discrete Time Wavelet Series

Besides discretizing the scale-translation plane, the independent variable (time or space) can also be discretized. A sequence of points (numbers) can be represented with a wavelet transform as well, and this wavelet transform is called the *discrete time wavelet series (DTWS)*. The DTWS is analogous to the discrete time Fourier series - both are discrete in the independent variable and the transform variable(s). A summation will replace integration in the wavelet transform. These DTWS's are analogous to the discrete time Fourier series that are implemented with Fast Fourier Transforms (FFTs).

The discrete time wavelet series is defined with respect to a "discrete mother wavelet", $h(k)$. The discrete time wavelet series maps $\mathbf{W}_h : l^2(\mathbf{Z}) \rightarrow l^2(\mathbf{Z}^2)$ (discrete finite energy sequences to a 2-D discrete grid of coefficients - Z represents the set of all positive, negative and zero integers). The **discrete time wavelet series (DTWS)** is:

$$\mathbf{W}_h f(m,n) = \frac{1}{\sqrt{a_o^m}} \sum_{k=-\infty}^{\infty} f(k)\, h\left(\frac{k-nb_o a_o^m}{a_o^m}\right) = a_o^{\frac{-m}{2}} \sum_{k=-\infty}^{\infty} f(k)\, h(a_o^{-m}k-nb_o)$$

$$= \langle f, h_{m,n} \rangle = \langle f, U(a_o^m, nb_o a_o^m)\, h \rangle \tag{2.33}$$

Note a critical point: the "discrete mother wavelet", $h(k)$, is evaluated at an argument that may not be an integer (it depends upon the values of a_o and b_o). Since the mother wavelet is a sequence, it is not defined for non-integer arguments. Obviously, one assumption that could be made is that both a_o and b_o are integers; however, the smallest non-unity integer is 2, so that the scale would be required to change by a minimum of a factor of 2. This dyadic scaling is drastic. Physical processes, such as the reflection process discussed in Chapter 1, very rarely scale a signal by any scale remotely close to 2. Finer scales are required to represent many physical processes. Also note that the mother wavelet is now denoted by "h" rather than "g"; this notation is consistent with filtering theory.

Under an appropriate set of assumptions (primarily bandlimited functions), mathematical tools such as interpolation can be applied to achieve finer scaling. Multirate filter banks achieve finer scaling (scales can be rational numbers) by passing the discrete signals through filters, interpolating, filtering again, and decimating [Vai1] but these structures can become extremely inefficient for many fine scale values (the efficiency of the multirate filter is due to it cascade structure - if many prime number filters are required to achieve the desired scaling, then the cascade structure will not be efficient). Interpolation, sampling theory, and approximation theory [Pap2 and many other uncited references] all investigate techniques that can be applied to achieve the finer sampling. This book does not investigate these different approaches but does assume that the discrete mother wavelet can be evaluated at non-integer values. However, further discussion of the

DTWS in this chapter will only address special wavelet transforms that only allow scales that are powers of 2 and integer translations.

Multiresolution, Orthogonal, and Biorthogonal Wavelet Transforms (and PR-QMFs)

Several other important wavelet transforms exist that are special cases of the continuous and discrete wavelet transforms (CTWS and DTWS, respectively). These wavelet transforms include multiresolution, orthogonal, biorthogonal wavelet transforms, and perfect reconstruction-quadrature mirror filters (PR-QMFs). The orthogonality, biorthogonality, and multiresolution properties can apply to continuous time signals as well as discrete signals but by using a discrete step size of 2 significant efficiencies can be achieved. When they are enforced for the discrete time signals, then special DTWS result. Each of these transforms will be defined and detailed in the following sections but this section addresses properties that are common to all of these transforms and is important to subsequent analysis.

For the multiresolution, orthogonal, biorthogonal, and PR-QMFs wavelet transforms, the mother wavelets must satisfy additional constraints, besides admissibility (and, in the discrete case, the lattice density and resolution requirements). To produce the orthogonality, multiresolution, and/or other properties, constraints are required to limit the set of mother wavelets to be smaller than just the set of admissible functions. These limitations are undesirable for some applications (e.g., representing systems that perform very fine scalings). However, for image representation, processing, and coding, these constrained wavelet transforms are ideal. For these applications the efficiency of the processing is as critical as the characteristics of the representation. However, for some image processing applications (i.e., one-to-many broadcast TV) significant processing can be committed to achieving better representations and, thus, alternative, less efficient computational wavelet transforms may be acceptable if they produce more efficient representations (fewer coefficients for the same degree of approximation). As with any model or tool, the application dictates which tools are efficient and which ones are not - a shovel is an efficient tool for prying loose dirt but not for loosening a tight hex nut.

Since these constrained wavelet transforms have already been extensively researched and published [Dau, Hei, Mal, Mey, Rio, Vet], the discussion in this book is brief and of a comparison nature. Initially (from the author's point of view), wavelet theory started in its nonorthogonal form with Morlet's modulated Gaussian mother wavelet. Obviously, extensive research was performed prior to this application, but the name "wavelet theory" and the broad interest really started from this application. As interest grew in this constant-Q transform, questions were posed regarding the existence of an orthogonal basis, and later interest arose in the existence of a compactly supported orthogonal basis. The orthogonal bases (even bases with compact support) were constructed. In related but parallel efforts, wavelet theory was significantly reformulated for discrete signals with the multiresolution wavelet transform. Multiresolution transforms were related to the pyramidal image decomposition scheme and perfect reconstruction-quadrature mirror filters (PR-QMFs) or subbanding. The PR-QMFs are related to biorthogonal wavelet transforms. Throughout this research many questions were answered affirmatively and were constructively resolved. In addition, the application of wavelet theory to images also flourished due to its significant efficiency gains. References to the detailed research is provided in the References section and this paragraph would require almost all of the references. The advantages of these constrained wavelet transforms briefly follows. The majority of the wavelet present wavelet theory research concentrates on these constrained wavelet transforms.

Although none of the constrained wavelet transforms has been defined yet, some of the commonalities between them are presented in this section. In their original forms most of them were formulated as discrete wavelet transforms with a scale step size of 2. For all of these transforms their mother wavelets were initially *real* and constrained (by enforcing a set of constraints these special transforms are constructed) although complex mother wavelets can be used. Before a detailed investigation of these transforms, a specific case of these transforms is investigated to illustrate the general structure of all of these transforms when applied to discrete data.

Discrete Time Wavelet Series - A Specific Structure

The discrete time wavelet series is discrete in both the time domain and the scale-translation (wavelet) domain. Since time dilation by a factor of 2 can be efficiently implemented simply by dropping every other sample of a discrete signal (decimating or subsampling by a factor of 2), the transform to be presented here only considers time scaling by powers of 2. The structure of this particular wavelet decomposition is presented in Figure 2.11. This structure can be valid for multiresolution, orthogonal, biorthogonal, or PR-QMFs, with each case specifying the requirements for the filter coefficients. The filters are simply denoted as low pass, $g(k)$, and high pass, $h(k)$, filters to present a general form, but these filters are intimately related to the mother wavelet of the wavelet transforms [Com, Dau, Mal, Mey, Rio, Vet]. The high pass filter, $h(k)$, is usually considered as the "mother wavelet" (the order of the coefficients changes in some cases) and the outputs of the high pass filters are thus the wavelet coefficients (the high pass filter convolves its impulse response (mother wavelet) with the incoming signal to create an output (a sequence of wavelet coefficients)).

Discrete Time Wavelet Series Block

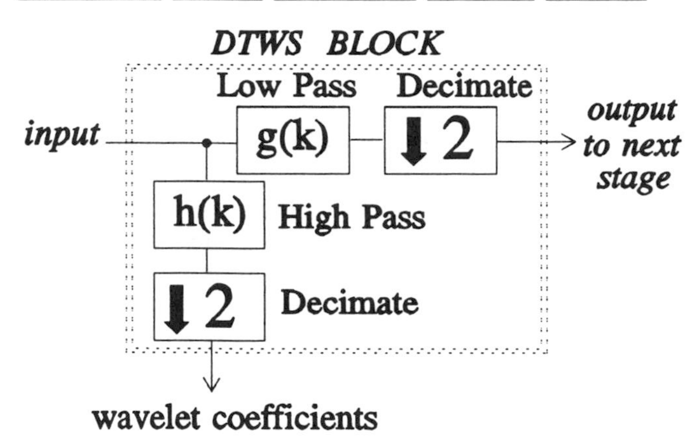

Figure 2.11: DTWS Processing Block

DTWS Decomposition

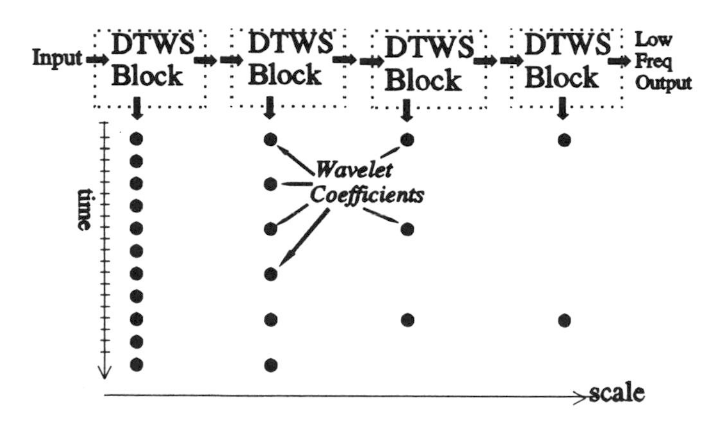

Figure 2.12: DTWS Structure

The decomposition process is demonstrated in Figure 2.12. The entire DTWS decomposition consists of passing the signal through identically structured processing "blocks." Each block is defined to have both a low pass filter (the "scaling function" discussed in wavelet literature) and a high pass filter (the "wavelet function"). The output of each filter is decimated by a factor of 2. The outputs of the lowpass filter are forwarded to the next DTWS block. The outputs of the high pass filters are the wavelet coefficients. These coefficients are the new representation of the signal. If the wavelet coefficients of Figure 2.12 are rotated 90 degrees clockwise, then the lattice structure will resemble time-scale or time-frequency resolution cells discussed in Figure 2.8; for these "band splitting" wavelet transforms the resolution cells of Figure 2.8 become appropriate.

Multiple "band splitting" blocks are concatenated to form a DTWS. The input signal goes in from the left, a series of wavelet coefficients comes out the bottom, and a final low frequency time series exits out the right. The low frequency signal and the wavelet coefficients together represent the time domain signal. This is one example of the DTWS. The "forward" transform is often termed the analysis filter or analysis stage. The pyramidal structure of this DTWS results because fewer and fewer coefficients are output from each successive stage until the last, single coefficient is output at the end of the filter stages (or the peak of the pyramid). The pyramidal decomposition is more easily viewed in the frequency domain and is discussed subsequently with multiresolution wavelet transforms (refer to Figure 2.15 to see the pyramidal decomposition in the frequency domain).

The corresponding inverse discrete time wavelet series (IDTWS) is displayed in Figure 2.13 and Figure 2.14. Its structure is identical to the DTWS but it operates in reverse. The IDTWS takes the low frequency sequence and the wavelet coefficients (high pass outputs) and reconstructs or synthesizes the original discrete time sequence. The subsampling or decimating by a factor of 2 is replaced by an upsampling process that interleaves a zero between each sample (effectively creating an interpolated signal after it is filtered). This IDTWS is often termed the synthesis stage. The properties of these low pass and high pass filters have not been detailed. The following sections provide some of the details of their design; however, the references [Dau, Rio, Vai, Vet] are a better source for details and examples.

Inverse Discrete Time Wavelet Series Block

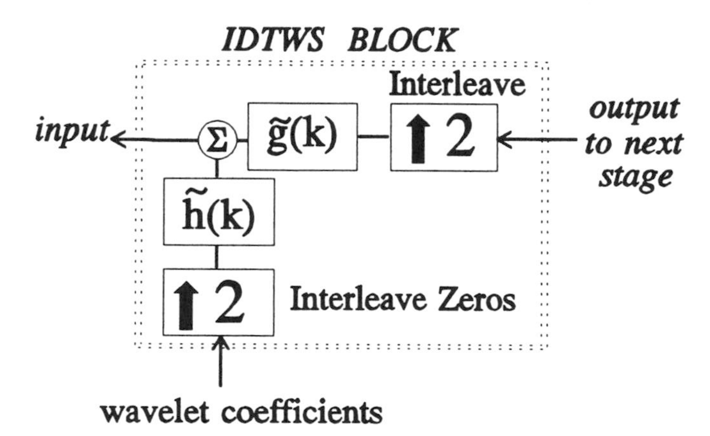

Figure 2.13: IDTWS Processing Block

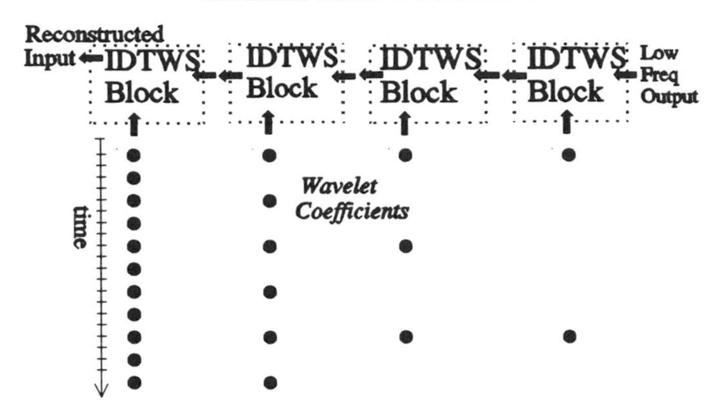

Figure 2.14: IDTWS Structure

Multiresolution Wavelet Transforms

From the list of special wavelet transforms, the multiresolution wavelet transforms are the most general [Mal, Mey]. Multiresolution wavelet transforms allow the mother wavelets to be nonorthogonal and have many other properties. The primary constraint on the mother wavelet (or high pass filter) is really formulated on a different function, the *scaling function* (or low pass filter). Multiresolution transforms "build-in" a pyramidal structure that is not required for general wavelet transforms. The pyramidal structure requires a repetitive application of the same (but scaled) scaling and wavelet functions, or lowpass and highpass filters, respectively. This pyramidal structure forces the scaling function to satisfy a constraint termed the two-scale equation (originally [Mal] and detailed in [Chu]). In addition, the multiresolution wavelet transforms often begin with a scaling function that is derived from a **spline** function (splines are usually simple functions, such as polynomials that can be efficiently represented). Many desirable advantages exist for using splines to derive mother wavelets (and these are detailed by Chui [Chu]) but the constraints imposed on the mother wavelet (the filters) limits the set of possible mother wavelets. Further details of the mathematics are deferred to the many references on multiresolution wavelet transforms [Chu, Com, Dau, Mal, Mey, Vet].

However, the standard application of the multiresolution wavelet transform is to form a series of half-band filters that divide a spectrum into a high frequency band and a low frequency band. These filters initially act on the entire signal bandwidth and, thus, act at the high frequencies (small scale values) first and gradually reduce the signal bandwidth with each stage (smaller bandwidths correspond to larger scales). See Figure 2.15. The high frequency band output is taken as the wavelet transform coefficients for a "fine" scale, and the low frequency band output is *decimated* by a factor of 2 (every other sample is discarded). This low frequency band is then split into a high and low band; this band splitting and decimation process continues and produces an octave band representation of the signal. The wavelet coefficients for a particular scale are the output samples of a particular high pass filter. Each different output sample corresponds to a different translation at that particular scale. The output rate of each of these filters is decimated by a factor of two as the scale value steps in the coarse direction. The pyramidal structure in Figure 2.15 results from the recursive structure of the multiresolution wavelet transform. The high pass filter outputs (wavelet coefficients) represent the signal's characteristics and energy at a particular scale. The output of the final lowpass filter is the residual or "d.c." portion of the signal - the most blurred (most coarse) signal.

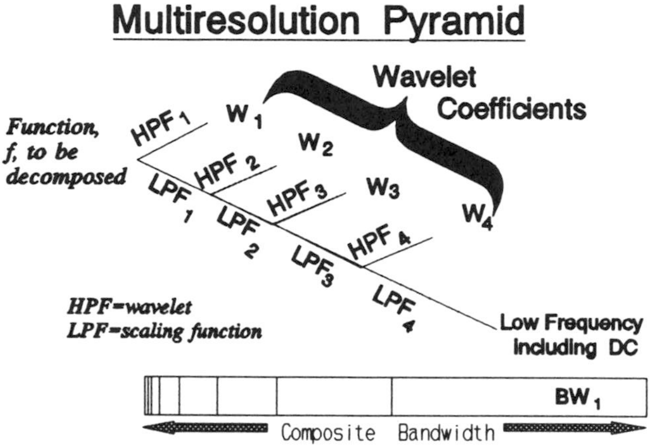

Figure 2.15: Multiresolution Wavelet Transform Structure

For this interpretation of the multiresolution wavelet transform it would be confusing if the mother wavelet was sophisticated (a large bandwidth or quickly changing characteristics); a frequency "band" may not make sense. If a sophisticated mother wavelet was used, the filters may have very sharp peaks or have multi-modal shapes; these are not acceptable in the standard multiresolution analysis. Thus, as with the other constrained wavelet transforms, the mother wavelets are assumed to be "bandpass" and not be sophisticated. The "sophisticated" mother wavelets to be examined later, are not necessarily characterized as bandpass. Subsequent ambiguity analysis will provide insight into the characteristics of these sophisticated mother wavelets.

In some practical applications (especially images) the signal information does appear to follow an octave band distribution, in many other applications the information does not (signal analysis). The pyramidal structure can become undesirable for applications in which the scale changes by very fine steps or applications in which the low frequency information becomes less important (system modelling as detailed in Chapter 5).

Orthogonal and Biorthogonal Wavelet Transforms

Just as in Fourier analysis and several other transform techniques, many desirable properties can be established if an orthogonal basis can be formed. Orthogonal projections allow an input function to be decomposed into a set of independent coefficients with a coefficient corresponding to each of the orthogonal basis elements. These independent or uncoupled coefficients are found by projecting the input function onto each basis elements (the projection is an inner product between the input signal and the basis elements). The orthogonality *decouples* each of the orthogonal basis elements and coefficients. Thus, each orthogonal wavelet coefficient represents its own piece of information about the signal - no redundancy exists in the representation. This leads to efficient representations and easy hypothesis testing (each hypothesis is independent of any others).

Orthogonal and biorthogonal bases were found or created for other transforms and, thanks to the work of many [Bat, Chu, Dau, Mal, Mey, Rio, Vet], orthogonal and biorthogonal bases have been constructed for the wavelet transform. Both orthogonal and biorthogonal wavelets are typically employed in the multiresolution wavelet decompositions. These techniques have been applied extensively in image compression, analysis and subband coding [Rio, Vai, Vet].

Orthogonal and biorthogonal wavelet transforms impose constraints on the mother wavelets (and/or scaling functions) so that the scaled and translated mother wavelets are orthogonal or biorthogonal. In the multiresolution scheme the orthogonality conditions are also enforced on the scaling function or lowpass filter. Recall that the orthogonality is enforced only at the discrete lattice points indexed by (m,n) ; the scale and translation step sizes, a_o and b_o, define the lattice points. Again, the majority of the research uses a scale step size of 2 and derives the conditions for this particular step size only. The **orthogonal wavelets** satisfy the condition that the inner product of the scaled and translated mother wavelet is an impulse in both scale and translation:

$$\int_{-\infty}^{\infty} g_{m,n}(t)\, g^*_{m',n'}(t)\, dt = \delta(m-m')\,\delta(n-n') = \begin{cases} 1 & \text{if } m=m' \text{ and } n=n' \\ 0 & \text{otherwise} \end{cases} \tag{2.34}$$

The orthogonal wavelet transforms have the desirable feature that the analysis and synthesis filters are the same (the high pass and low pass filters in Figure 2.11 and Figure 2.13 are identical). The details for the discrete implementation are well covered in [Vet1].

Biorthogonal wavelet transforms are closely related to subband coding [Vai, Vet]. These biorthogonal transforms require four different filters (the high pass and low pass filters in Figure 2.11 and Figure 2.13). These filters must satisfy the biorthogonality constraints. The biorthogonal constraints essentially state that the combination of the analysis and synthesis stages should be an identity operation. When compared to orthogonal wavelet transforms, biorthogonal transforms relax some of the constraints on the mother wavelet (or filters) and allows the mother wavelets to be symmetric and have linear phase.

Since the biorthogonal wavelet transforms are supposed to be perfect reconstruction transforms, no information should be lost or filtered out. Thus, the decomposed signal at the input to the analysis stage should be identical to the reconstructed output of the synthesis stage. Intuitively this states that the overall system should be an identity system. However, an additional constraint is the "orthogonality." The orthogonality condition states that the representation should not have any redundancy. The zero redundancy requires that the information in one branch of the decomposition be independent of the information in the other branch of the decomposition (since there are two branches in the basic "block," this condition is referred to as bi-orthogonality. How can both perfect reconstruction and biorthogonality be enforced? The overall system must be the identity system and each branch must have independent information. The filters must satisfy these conditions. Denote each filtering action simply by its appropriate capital letter (G represents g(k), H represents h(k), etc). Considering only a single block of the decomposition (as in Figure 2.11 and Figure 2.13) and combining them to form an overall system is shown in Figure 2.16.

Single Stage Analysis/Synthesis: Overall System

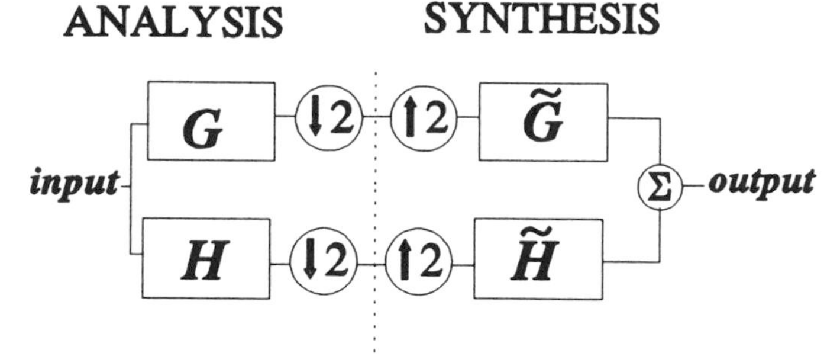

Normally some type of coding occurs in between these stages

Figure 2.16: Single Stage PF-QMF

The biorthogonality constraints (these are the same conditions enforced for PF-QMFs) on these filters first require that the overall system be an identity system (perfect reconstruction) or that:

$$G\tilde{G} + H\tilde{H} = I \qquad (2.35)$$

The second biorthogonality constraint requires that each channel or branch have independent information or that no redundant information is filtered out. In terms of the filters this orthogonality condition requires:

$$G\widetilde{H} = 0 \quad and \quad H\widetilde{G} = 0 \tag{3.36}$$

These conditions can be imposed directly onto the filter coefficients for implementation purposes and this is done in [Vet1]. The biorthogonal and multiresolution wavelet conditions were imposed on discrete time signals with a scale step size of 2. The resulting discrete time filters are the perfect reconstruction-quadrature mirror filters (PR-QMFs) [Rio, Vet].

Besides the multiresolution, orthogonal, and biorthogonal wavelet constraints, some wavelets are constrained to be regular and compactly supported (or smooth and nonzero only on a finite interval) as well. And, as mentioned previously, Daubechies has combined all of these constraints to produce *unique* wavelets that have taken her name, the Daubechies' wavelets [Dau].

The constraint that the mother wavelet is real-valued is significant for system formulations and imaging applications. If the mother wavelet is real, then, just as in a narrowband real receiver, the phase will modulate the magnitude of the output. The surface created by the magnitude of the receiver output will oscillate rather than being smooth as in Figure 2.10. This was demonstrated in the wavelet transform examples presented earlier in this chapter. The phase and magnitude are inseparable and lead to interference phase patterns on the magnitude of the wavelet transform. For many applications the real mother wavelet constraint is undesirable and an analytic (complex) mother wavelet is preferred. Requiring that the mother wavelet is analytic may invalidate some orthogonal/biorthogonal wavelet constructions and more general constructions or constraints may be required.

Further constraints on the mother wavelets will not be considered. The constraints on the mother wavelets are undesirable for some applications such as the time-varying system theory considered in Chapter 5. The constraints also modify and, in some cases, limit the resolution properties of the wavelet transforms. The resolution properties are considered in the next chapter.

Before leaving the biorthogonal wavelet transforms, reconsider Figure 2.16. The figure has a dashed line dividing the analysis stage from the synthesis stage. It is noted on the figure that "coding" is often applied in between these two stages. As mentioned at the start of Chapter 1, speech is often recognized by breaking a phrase or sentence down into recognizable pieces or words and then reconstructing the phrase or sentence from these pieces. "Coding" is the formal application and study of this concept. By putting a message into a redundant framework (encoding the message) then the coded message can suffer distortion or have noise added to it but by properly deciphering the distorted signal, the original message can be recovered. Often the first step to encoding a message is to break the original message into pieces and encode each piece. This is the analysis stage.

Besides the application of "direct" coding as discussed in the previous paragraph, the same concept can be applied to achieve data compression or more efficient transmission. Consider image processing as an example. If an image is to be sent from one point to another (like a FAX machine) then the image must be encoded to be sent over the transmission medium (phone lines). By using an analysis filter bank the image can be decomposed into separate channels (subbands). Suppose some of the channels had very little energy in them while several other channels had the majority of the signal energy. If signal was broken into 8 bands with the same data rate but only 4 bands could be sent over the transmission medium, then the 4 bands with the most energy could be sent. Thus the transmitted energy might be a good approximation to the original image but consist of only half of the original "size." This example is an extremely simple example of image coding and quantization (voice and other signals can also be considered) and much better examples and more detail can be found in the signal and image processing literature and journals. However, this is another application of the wavelet decomposition process and one that is well suited to the orthogonal, multiresolution, and biorthogonal wavelet transforms.

An Image Processing Example

As mentioned several times, the multiresolution, orthogonal and biorthogonal wavelet transforms are very efficient for image analysis. To briefly demonstrate the application of these wavelet transforms to image analysis and representation, consider the following example. The image of Figure 2.17 will be decomposed by a multiresolution wavelet transform. This is an unjustly summarized example that follows the research presented in [Dau, Mal, Mey, Rio and, more recently, many others].

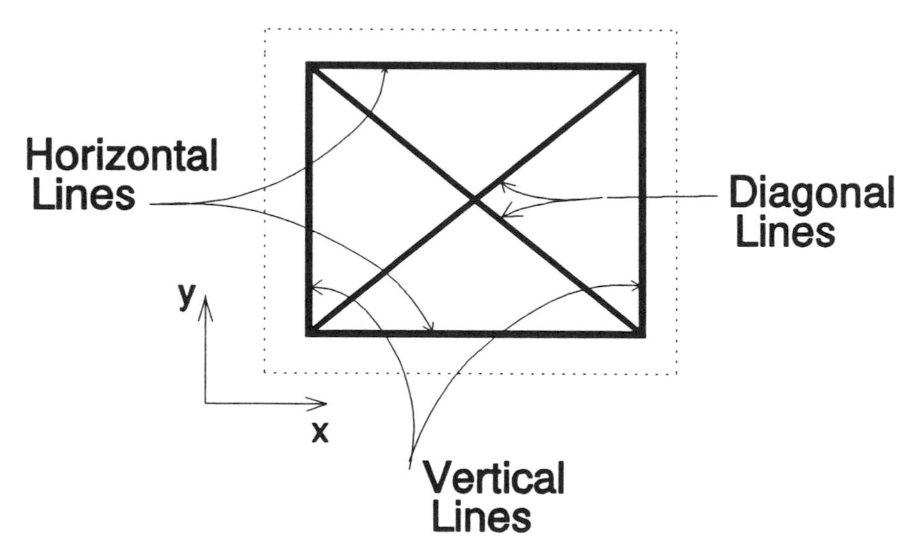

Figure 2.17: Image Example

The image consists of 6 lines, two vertical, two horizontal and two diagonal. The rest of the image is assumed to be zero. Only the region inside the dashed line is being decomposed (note that between the edge of the figure (dashed lines) and any of the 6 lines, a "buffer zone" of zeros exists - this "buffer zone" of zeros eliminates edge effects and filter initializations that are not discussed here).

This image will be decomposed with the multiresolution wavelet transform discussed in the previous sections. This multiresolution wavelet transform must be extended from operating on one-dimensional signals (such as a time series) to operate on two-dimensional images. However, for the sake of simplicity, the image will be broken down into two "separable" dimensions [Bos]. The separability of the image coordinates allows the image to be broken down into a series of one-dimensional sequences and, thus, allows the previously discussed one-dimensional filtering to be applied. The two-dimensional image is mapped into two one-dimensional sequences. This mapping or re-ordering of the image pixels is termed lexicographical ordering. The first one-dimensional sequence (or the horizontal sequence) is formed by taking the first (top) row of the image (starting at the pixel in the upper left-hand corner of the image and using this pixel as the first point in the one-dimensional sequence). Scanning along this top horizontal row of the picture creates the ordered sequence. The scanning is then repeated for the other rows of the image to create a series of one-dimensional sequences (however, all of these horizontal sequences are treated identically, so the rest of the processing just refers to "the" horizontal sequence). These one-dimensional sequences are then wavelet transformed to compute the (horizontal or x-axis) wavelet domain coefficients.

This same scanning process is repeated in the vertical direction to create a series of vertical or y-axis one-dimensional sequences. These sequences are similarly wavelet transformed. Recall that the standard multiresolution configuration employed a scale step size of 2 (which is really just subsampling) and essentially splits the one-dimensional sequences into a highpass band and a lowpass band. This band splitting operation is applied to both the horizontal and vertical one-dimensional sequences in cascade (see Figure 2.18).
After the horizontal filters and the subsampling the one-dimensional sequences (filter outputs) are placed back into two-dimensional images (sub-images). These images are then scanned to create the vertical sequences. These vertical sequences are then filtered and subsampled and the resulting sequences are again returned to the two-dimensional image format; these are the four sub-images of the original image. These sub-images are the wavelet domain representation.

In applying the same wavelet transform separably to both the horizontal and vertical sequences [Mal, Mey] it was assumed that the scaling functions and mother wavelets were also separable. More general non-separable multidimensional wavelet transforms are being developed and applied and are the subject of current research.

Returning to the example, recall the operation of a lowpass filter. If a signal with low frequency content (slow variation) is input into the lowpass filter, then the response (output) of the filter is large relative to the filter response to a signal with only high frequency content (fast variations). For the image in Figure 2.17, a single horizontal line will yield a sequence of consecutive numbers that are all the same for the horizontal sequence (a lowpass signal). But, for the vertical sequence, this same horizontal line yields an impulse and then many consecutive zeros, thus, the horizontal line is a highpass signal for the vertical

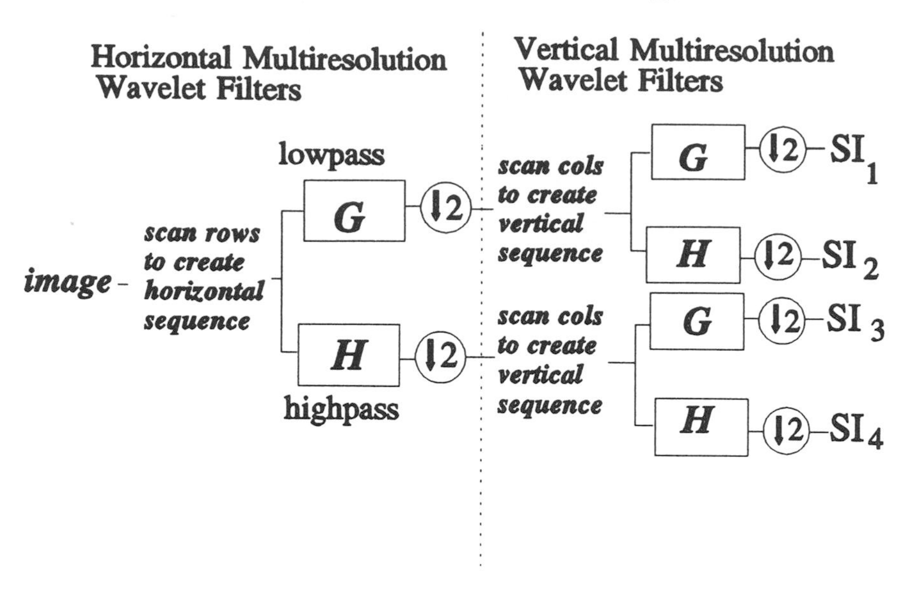

Figure 2.18: One Stage of a Separable Wavelet Decomposition of an Image

sequence. The reverse occurs for a vertical line (highpass in the horizontal sequence and lowpass in the vertical sequence).

Now consider one of the diagonal lines. In both the horizontal and vertical one-dimensional sequences the diagonal line appears as a series of impulses. Thus, the diagonal line is a highpass signal in both the horizontal and vertical directions. Consider an image feature that is not present in Figure 2.17, a solid section (rectangle, square, circle, etc.). A solid will be "unwrapped" by the lexicographical ordering of the horizontal and vertical sequences. Both of these resulting sequences will have segments that are constant over consecutive points (low frequency components). Therefore, a solid will be lowpass in both the horizontal and vertical sequences (although the edges will be extracted as highpass components).

As in the one-dimensional case, the high frequency components are considered as the wavelet coefficients and are part of the final wavelet domain representation. The high frequency components are considered as any of the images that passed through either or both of the highpass filters. For the image example, three such wavelet domain sub-images are created (often called the "detail" sub-images); the horizontal and vertical highpass sub-image (diagonal lines), the horizontal highpass-vertical lowpass sub-image (vertical lines), and the horizontal lowpass-vertical highpass (horizontal lines). The remaining lowpass-lowpass sub-image would ordinarily be passed through another bank of four filters to further decompose the image (creating a "pyramidal decomposition" [Bur, Mal]). But for this particular image, only highpass components are present and the lowpass-lowpass sub-image is essentially zero everywhere and so it is not decomposed any further.

The multiresolution wavelet decomposition of the image in Figure 2.17 is shown in Figure 2.19. The four sub-images are labeled. This image decomposition was performed with respect to an extremely simple mother wavelet, the real Morlet mother wavelet with only about three significant cycles. Recall from an example earlier in Chapter 2 that the wavelet transform of an impulse is just scaled versions of the mother wavelet translated across the scale-translation plane. Thus, the response of the highpass filters due to the impulses will be just the mother wavelet. Instead of showing the mother wavelet as the output of these filters, a very simple image thresholding causes these sharp mother wavelets to just be points. Therefore, the highpass outputs return the original sharp lines instead of versions that are blurred by the mother wavelet. Note how, again, the mother wavelet controls the resolution characteristics of the wavelet decomposition.

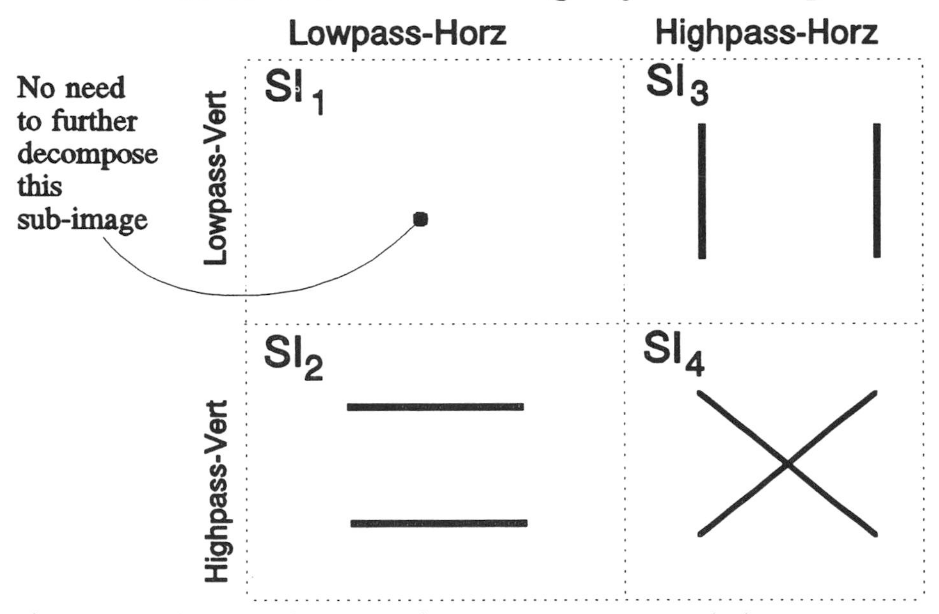

Figure 2.19: Multiresolution Image Decomposition

Due to the subsampling of each filter output the four sub-images each only have one fourth of the points as the original image. An efficiency of this wavelet decomposition should be immediately clear. The lowpass-lowpass sub-image can be discarded without losing any of the information about the image - that particular "subband" does not need to be transmitted. Besides this trivial efficiency improvement, the implementation of the wavelet or subband filters is also very efficient [Com, Mal, She].

For simplicity the image in this example did not include any solids (lowpass signals in both the horizontal and vertical directions). However, in real images the lowpass-lowpass information is often the most valuable for conveying information. The lowpass-lowpass portion of the image is just a "blurred" version of the original image (analogous to looking at something far away through a pair of glasses that causes nearsightedness); most of the information in the image can still be extracted. Therefore, in coding real images, this lowpass-lowpass image information is often emphasized and given the highest priority (unlike the de-emphasis in the particular image that was considered).

This section presented a brief and introductory example of the application of wavelet theory to image analysis and coding. Much better and more detailed examples can be found in the references, especially [Mal].

Appendix 2-A: Nonunique Wavelet Domain Representations

This appendix demonstrates that the wavelet domain representation of a signal is not unique *even with respect to the same mother wavelet.* Different wavelet domain distributions can be inverse transformed to reconstruct the same time (or space) domain signal. Or, stated otherwise, multiple wavelet domain representations for the same time domain signal exist. Refer ahead to Figure 2.20-Figure 2.22 for a visual demonstration that is subsequently analyzed. The wavelet transform is a one-to-one operator - only one wavelet transform (wavelet domain distribution) exists for a given signal. The inverse wavelet transform always maps a specific wavelet domain representation back to a specific time domain signal as well. However, multiple wavelet domain representations can lead to the same time domain signal being reconstructed.

Consider the inverse wavelet transform of a wavelet transform:

$$h(x) \doteq \mathbf{W}_g^{-1}[\mathbf{W}_g\{f(x)\}] = f(x) \tag{A.1}$$

For this case the inverse wavelet transform of the wavelet transform of f is always f or $h(x) = f(x)$. It is unique. Switching the order of the "forward" and inverse wavelet transform yields:

$$H(a,b) \doteq \mathbf{W}_g[\mathbf{W}_g^{-1}\{F(a,b)\}] \tag{A.2}$$

<u>The new wavelet transform representation</u>, $H(a,b)$, <u>can be different than the original wavelet domain representation</u>, $F(a,b)$. In general, *multiple wavelet domain representations can represent the same time function, even with respect to the same mother wavelet.* Writing out the previous equation, and substituting s and τ for a and b, respectively, in the wavelet transform yields:

$$H(s,\tau) \doteq \int_{-\infty}^{\infty} \left[\frac{1}{c_g} \int_{-\infty}^{\infty}\int_{-\infty}^{\infty} F(a,b) \frac{1}{\sqrt{|a|}} g\left(\frac{x-b}{a}\right) \frac{da\, db}{a^2} \right] \frac{1}{\sqrt{|s|}} g^*\left(\frac{x-\tau}{s}\right) dx \tag{A.3}$$

or, by reordering the integration:

$$H(s,\tau) \doteq \frac{1}{c_g} \int_{-\infty}^{\infty}\int_{-\infty}^{\infty} F(a,b) \left[\int_{-\infty}^{\infty} \frac{1}{\sqrt{|a|}} g\left(\frac{x-b}{a}\right) \frac{1}{\sqrt{|s|}} g^*\left(\frac{x-\tau}{s}\right) dx \right] \frac{da\, db}{a^2} \tag{A.4}$$

$$= \frac{1}{c_g} \int_{-\infty}^{\infty}\int_{-\infty}^{\infty} F(a,b) \left[\mathbf{W}_g\, [g(x)] \left(\frac{s}{a}, \frac{\tau-b}{a}\right) \right] \frac{da\, db}{a^2} \tag{A.5}$$

This equation is referred to as the *reproducing kernel equation* and will be subsequently detailed. The nonuniqueness condition states that several $F(a,b)$'s will lead to the same $H(s,\tau)$. If an inverse wavelet transform of $H(s,\tau)$ is then taken, it will always be the same $h(x)$. So different $F(a,b)$'s will have produced the same $h(x)$.

Nonuniqueness Example

Although the analysis in this section is abstract and can be skipped on an initial reading, it is important for the system identification or scattering estimation in the wavelet domain (or for wideband/nonstationary signals and space-time-varying systems - the application of this analysis is detailed in Chapter 6).

A property of wavelet transforms is that the reconstructed signal will not change if the wavelet domain representation is "smeared" by the wavelet transform of the mother wavelet, $\mathbf{W}_g g(a,b)$. *The wavelet transform of the mother wavelet,* g*, (with respect to the mother wavelet) is thus called the* ***reproducing kernel*** *since the same wavelet transform is reconstructed (reproduced) from the original wavelet transform* [Com].

$$\mathbf{W}_g f(a,b) = \int_{-\infty}^{\infty}\int_{-\infty}^{\infty} \mathbf{W}_g f(a',b')\, \mathbf{W}_g g\left(\frac{a}{a'}, \frac{b-b'}{a'}\right) \frac{da'\, db'}{(a')^2} \tag{A.6}$$

The kernel of this integral operator is the wavelet transform of the mother wavelet; the reproducing kernel becomes a natural name.

The impulse function acts similar to a reproducing kernel for the convolution operator. The sifting property for the convolution operator provides a reproducing kernel property; if the impulse is convolved with any other function it reproduces that function as the output [Bos, Kuo]. In equation form:

$$f(t)*h(t)=\int_{-\infty}^{\infty}f(\tau)h(t-\tau)d\tau=\int_{-\infty}^{\infty}f(\tau)\delta(t-\tau)d\tau=f(t) \qquad \text{(A.7)}$$

For equation (A.6) the operator is the wavelet transform instead of the convolution operator, and the reproducing kernel is the wavelet transform of the mother wavelet instead of the impulse function.

Reproducing kernels are important for the practical application of system identification. The goal of the inverse system estimation problem (system identification) is to construct a system model of the real system (an estimate of the system). Taking the "inverse" of the system model creates an inverse system model. By creating a composite system that is a cascade of the *real* system and the inverse system model, a measure of the accuracy of the system model can be achieved. If the inverse system model was perfect, then the composite system would be just a pass through system (identity system). For a given input, the output of this composite system should equal the input. The performance measure is created by examining the differences between the composite system's input and output. Since the composite system should recreate the input at the output, it acts just like a reproducing kernel and its associated operator.

Consider the convolution example - if the composite system's impulse response is just an impulse (the reproducing kernel for the convolution operator), then a good system model was created. Similarly, for the wavelet model, if the composite system response is the reproducing kernel, then a good system model was created. The extra freedom involved in the mother wavelet does not allow the reproducing kernel to be known a priori. The model of the composite system should just be the reproducing kernel. Thus, another performance measure (in a different domain) is available.

Consider a specific example applicable to wavelet transforms. Two different wavelet domain distributions are considered: the first wavelet domain representation is the wavelet transform of the mother wavelet that is centered at a particular scale and translation value; the second representation is an impulse function (zero everywhere except at one point) at the same scale-translation value as the first representation. See Figure 2.21-Figure 2.22. This important example is the nonuniqueness of the wavelet domain representation of the mother wavelet, $g(t)$, itself. Taking the forward transform of any mother wavelet, $g(t)$, with respect to itself yields the first wavelet domain representation of $g(t)$ (F_1 denotes the first wavelet domain representation):

$$W_g[g(t)](a,b)=\int_{-\infty}^{\infty}g(t)\,\frac{1}{\sqrt{|a|}}\,g^*\left(\frac{t-b}{a}\right)dt \qquad \text{(A.8)}$$

$$=F_1(a,b)$$

This is the first representation of the mother wavelet and it is shown in Figure 2.20 for a specific mother wavelet (magnitude only). This transform domain representation will have support over many scale and translation values - since this equation is valid for any admissible mother wavelet, even the so-called orthogonal wavelets will have support over many scales and translations because they are only orthogonal on a specific grid of scale-translation values. This first wavelet domain representation is labelled $F_1(a,b)$.

Returning to equation (A.6) and rewriting it with $f=g$ and $(a,b)=(s,\tau)$ yields:

$$\begin{aligned} W_g g(s,\tau) &= \frac{1}{c_g} \int_{-\infty}^{\infty}\int_{-\infty}^{\infty} W_g g(a,b)\, W_g g\left(\frac{s}{a}, \frac{\tau-b}{a}\right) \frac{da\, db}{a^2} \\ &= \frac{1}{c_g} \int_{-\infty}^{\infty}\int_{-\infty}^{\infty} F_1(a,b)\, W_g g\left(\frac{s}{a}, \frac{\tau-b}{a}\right) \frac{da\, db}{a^2} \end{aligned} \tag{A.9}$$

Observe from that the same (which is $W_g g(s,\tau)$) can be obtained from a second, different, wavelet domain representation, $F_2(a,b) = c_g\,\delta(a-1)\,\delta(b)$:

$$\begin{aligned} &\frac{1}{c_g} \int_{-\infty}^{\infty}\int_{-\infty}^{\infty} F_2(a,b)\, W_g g\left(\frac{s}{a}, \frac{\tau-b}{a}\right) \frac{da\, db}{a^2} \\ &= \frac{1}{c_g} \int_{-\infty}^{\infty}\int_{-\infty}^{\infty} c_g\,\delta(a-1)\,\delta(b)\, W_g g\left(\frac{s}{a}, \frac{\tau-b}{a}\right) \frac{da\, db}{a^2} \\ &= W_g g(s,\tau) \end{aligned} \tag{A.10}$$

This is another valid wavelet domain representation of the time domain mother wavelet. This second representation is a delta function at (a=1, b=0). Thus, $F_1(a,b)$ and $F_2(a,b)$ are both equivalent wavelet domain representations of $g(t)$; the inverse wavelet transform of both $F_1(a,b)$ and $F_2(a,b)$ yield $g(t)$. Try it for any general $g(t)$. If the wavelet domain representation is $F_2(a,b) = c_g\,\delta(a-1)\,\delta(b)$, then:

$$W_g^{-1}[c_g\,\delta(a-1)\,\delta(b)] = \frac{1}{c_g} \int_{-\infty}^{\infty}\int_{-\infty}^{\infty} c_g\,\delta(a-1)\,\delta(b) \frac{1}{\sqrt{|a|}}\, g\left(\frac{t-b}{a}\right) \frac{da\, db}{a^2} \tag{A.11}$$

$$= \int_{-\infty}^{\infty} \delta(b) \int_{-\infty}^{\infty} \frac{\delta(a-1)}{|a|^{\frac{5}{2}}}\, g\left(\frac{t-b}{a}\right) da\, db = \int_{-\infty}^{\infty} \delta(b)\, g(t-b)\, db = g(t)$$

Again, if impulse functions are annoying to the reader, then limiting conditions can be utilized to construct a more rigorous, but identical result (a Gaussian function with its variance approaching zero in the limit approximates an impulse function).

The inverse wavelet transform of $W_g g(a,b)$ is, by the definitions, $g(t)$. It is critical to understand that $g(t)$ can be exactly recovered from *either* of these, nonequal, wavelet domain representations (see Figure 2.20-Figure 2.22 for a pictorial representation). These are two of the nonunique wavelet domain representations of the mother wavelet; other representations exist as well. However, it must also be noted that the wavelet transform defined in equation (2.2) only created the first representation, $W_g g(a,b)$, and the impulse representation originated elsewhere.

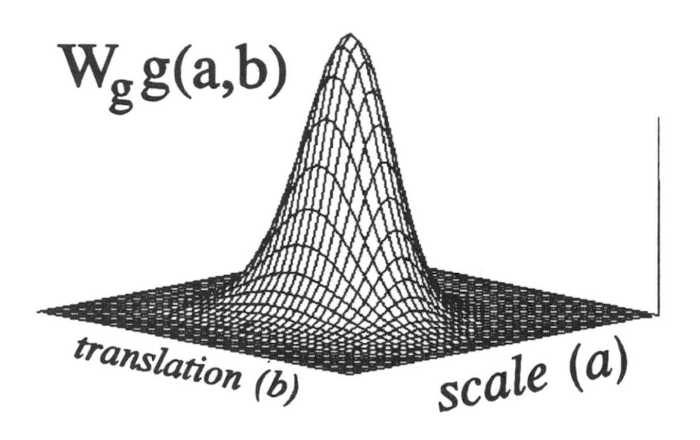

Figure 2.20: Magnitude of Wavelet Transform of Mother Wavelet

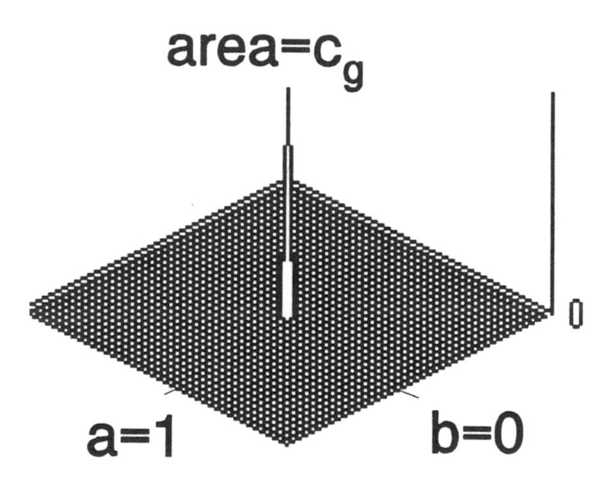

Figure 2.21: Wavelet Domain Representation of a Mother Wavelet

Nonunique Wavelet Domain Representation

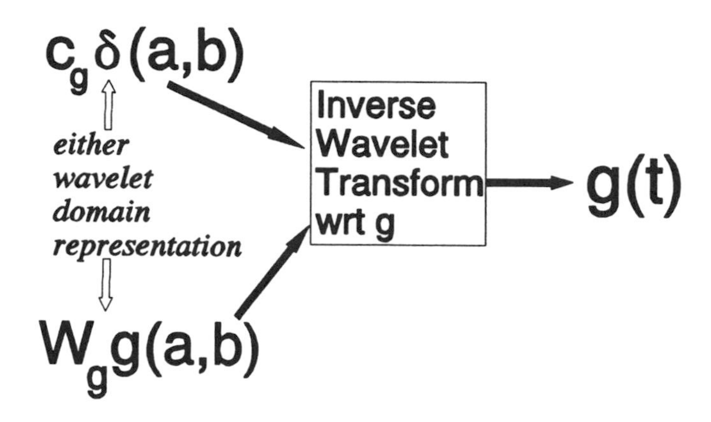

Figure 2.22: Nonuniqueness of Inverse Wavelet Transform

Although the impulse function may not be the minimum norm representation, it becomes the desired wavelet domain representation for "inverse problems" such as imaging or system identification (this is detailed in Chapter 6). The impulse function has good properties if the peak(s) of the wavelet surface is to be estimated or if the wavelet transform is buried in noise. The impulse model helps to visualize and simplify the concepts and mathematics and demonstrates a familiar example to those familiar with Fourier transforms and narrowband scattering or system identification [Kuo, Zio].

The nonuniqueness of the wavelet domain representation can sometimes lead to invalid conclusions and conflicting answers. When working with inverse wavelet transforms of any general distribution in the scale-translation plane, the nonuniqueness property should always be considered. Later, in Chapter 6, after wavelet domain system models are established, the nonuniqueness property becomes crucial. The scattering problem or channel modelling application cannot be properly addressed unless this nonuniqueness property is considered. This property arises in point scattering or point reflector models and is essential to properly interpret the physics, processing and mathematics jointly.

Chapter 3: Practical Resolution, Gain, and Processing Structures

Introduction

Applying wavelet theory in many scientific fields requires general design criteria or performance measures that can be traded off against one another. The tradeoffs between using wavelet transforms versus other transforms are usually stated in terms of resolutions, gains, robustness, noise immunity, and many other criteria. For wideband signals and systems the criteria are often measured with uncertainty functions, point-spread functions, and ambiguity functions.

Most signal processing and modelling applications employ narrowband or Fourier processing rather than wideband or wavelet processing; therefore, the narrowband and wideband performance measures and processing structures must be related and compared. Narrowband and wideband processing are compared in this chapter. This chapter also presents, analyzes, and compares several wideband or wavelet processing structures.

Although Fourier and wavelet processing have been compared and analyzed before [Com, Dau, Mal, Rio], a significant feature of this book is the investigation of the wideband or wavelet processing with *"sophisticated" mother wavelets*. A sophisticated signal has a time-bandwidth product that significantly exceeds one. Some non-sophisticated mother wavelets have been investigated (e.g., the Morlet Gaussian weighted tone mother wavelet as in Figure 1.1). The transforms for these non-sophisticated wavelets may be more computationally efficient; however, the gain, resolution, and other desirable properties may not be sufficient for some applications. This chapter examines and characterizes the effects (parameterization) of the mother wavelet on the resulting wavelet transform representation and identifies some of the advantages of using sophisticated mother wavelets.

Properties of the mother wavelet control many of the characteristics of the wavelet transform representation. The resolution properties of tonal-like mother wavelets have been thoroughly investigated and have been shown to outperform the Fourier transform for some applications. However, more sophisticated mother wavelets can also be employed. Due to the freedom in choosing the mother wavelet, each application may have its own, most efficient or useful, mother wavelet (or set of mother wavelets). Other applications may not know "good" mother wavelets a priori, and for those applications the mother wavelet can be changed, updated, or designed in real-time.

Besides the parameterization by the mother wavelet, the wavelet transform's correlation structure is exploited to relate wavelet transforms to well understood concepts. The wavelet transform has a correlator or template matching structure. Each wavelet domain coefficient, $W_g f(a,b)$, at a particular scale and translation (a,b) indicates how well the signal, f, matched the particular scaled and translated mother wavelet, $g_{a,b}$. *The "degree" of match or measure of closeness to the signal controls the magnitude of the wavelet coefficient or the gain of that particular scaled and translated mother wavelet.* This gain will be large if the signal being analyzed and the particular mother wavelet match. *However, the magnitude of this gain is also controlled by the time-bandwidth product (or "sophistication") of the chosen mother wavelet.* For a short tonal mother wavelet, even a perfect match to the signal may provide a relatively small gain. If the mother wavelet is a *sophisticated mother wavelet* (has a time-bandwidth product significantly greater than one), then the gain can be relatively large. For sampled data and a fixed time interval, many more samples are required for the sophisticated mother wavelet than for a non-sophisticated mother wavelet. Just as with any correlator or template matching structure, the more samples that are matched, the higher the gain. If variations of the signal or system can be tracked and characterized, then the modelling can be applied over a bigger interval to achieve higher gains or resolutions (than models that cannot track the variations).

Besides the gain of the representation, the resolution can be similarly related to the properties of the mother wavelet. If the gain of the representation drops quickly from the peak gain value(s), then the representation usually has good resolution, or a sharp peak has better resolution than a broad peak. The interpretation of good resolution is application dependent, so this book attempts to generally address the resolution properties by characterizing the resolution properties controlled by the mother wavelet. It will be shown that *the finest resolution that a wavelet transform can achieve depends upon the resolution properties of the mother wavelet at the scale origin (a=1).* The resolution properties of the mother wavelet are described by the uncertainty functions, point-spread functions, or ambiguity functions. Since the ambiguity function terminology is the mostly widely accepted of these functions (and to avoid repeating "uncertainty function, point-spread function, or ambiguity function") only the ambiguity function will be discussed.

Fourier/Narrowband Gain and Resolution Comparisons

Just as in any analysis, various performance measures must be applied to compare one technique to another. One of the primary techniques for comparing transform techniques is the resolution that they can achieve. For time-frequency or time-scale transform techniques such as the wavelet transform, the Fourier transform, the Gabor transform, Wigner analysis, and others, one performance measure of interest is the *resolution* in the appropriate domain. For wavelet transforms the resolution describes how well the translation (i.e., time delay or spatial shift) and the scale (corresponding dilation parameter) can be *simultaneously* resolved. In many applications both the resolution and the processing gain are important. The processing gain is often measured by the product of the bandwidth and the valid processing interval (time-bandwidth product).

Wavelet transforms are often compared to Fourier transforms due to their similarities and to emphasize some applications where wavelet theory should be advantageous over Fourier analysis. Fourier transforms, windowed Fourier transforms (FTs), Fourier series, short-time Fourier transforms (STFTs), discrete Fourier transforms (DFTs), and fast Fourier transforms (FFTs) are discussed in many reference texts [Bos1, Hla, Pap2, Rio, Van] and are not repeated in this book. The considerations of continuous time and discrete time signals should also be fully understood to appreciate the following analysis.

Before detailing the specific resolution and gain properties associated with wavelet transforms, consider the structure of the Fourier and wavelet transforms. The Fourier transform maps a one dimensional function of time to a one dimensional function of frequency - a weighted sum of tones with the weights being the Fourier coefficients. The short-time Fourier transform (STFT) maps a function to a weighted sum of windowed tones with a fixed window size. Many researchers have identified that the wavelet transform (with a tone-like mother wavelet) maps a function analogous to the STFT that has *a changing window size* [Dau, Rio]. As discussed in Chapter 2, the window size changes to maintain a constant ratio between the center frequency and the bandwidth (constant-Q) of the scaled mother wavelet. However, the kernel of the wavelet transform (the mother wavelet) does not need to be a windowed tone; it could be a windowed frequency modulated (FM) signal or several FMs existing simultaneously. Then the wavelet characterization is harder to intuitively compare to the Fourier techniques - but it should be compared, and this comparison is achieved by utilizing ambiguity functions. The freedom of using admissible functions as the kernel of the transform (and not just tones or windowed tones as in the Fourier transform) is extremely significant and is of primary importance to applications considered later in this book.

The resolution characteristics of the wavelet transform representation can be demonstrated in the time-scale plane for specific mother wavelets. Examples of the resolution are shown in Figure 2.8 and Figure 3.3. Recall that these figures represent the resolution of the wavelet transform in characterizing mother wavelets; as stated previously, the finest achievable resolution is dictated by the wavelet transform of the mother wavelet at the scale origin (a=1). The scale origin condition is required because the signal's resolution changes with scale (unlike the narrowband resolution as shown in Figure 3.2). The scale parameter controls the duration of the signal; if a short windowed tone (Morlet mother wavelet) is scaled to stretch it into a longer signal, then its resolution properties in the time-frequency or time-scale plane also change. Thus, the scale origin must be chosen as the "resolution reference" for wavelet transforms.

Each resolution cell represents a wavelet transform *of a mother wavelet* that is scaled and translated to a particular scale-translation value (a', b'), with respect to that same mother wavelet, $\boldsymbol{W}_g\left[g\left(\frac{x-b'}{a'}\right)\right](a,b)$. A single resolution cell for a tone-like mother wavelet is shown in Figure 2.10. The resolution at a different scale is determined by wavelet transforming a mother wavelet that has been warped by the new scale value. As can be seen from Figure 2.8, the resolution over the entire plane includes wavelet transforms of many scaled mother wavelets. Notice that the resolution cell size does not change due to changes in translation.

Consider a specific example. The time domain signal in Figure 3.1 is a sum of a short duration, high frequency transient windowed tone (nearly a delta or impulse function) and a long duration, lower frequency windowed tone. The two components are easily resolved by a wavelet transform with a Morlet mother wavelet as shown by the magnitude plot also in Figure 3.1. A STFT performed on this same signal will result in the resolution of only one of these tones. With a very short window, the STFT will localize the short tone, while with a long window, the STFT will only localize the long tone. The STFT resolution is controlled by the chosen window size and the window size smears one of the tones; the simultaneous resolution of both tones is difficult to achieve with standard Fourier techniques. When speech signals are analyzed with STFTs, several different window sizes are often used to analyze different features of the speech (i.e., distinguish a relatively long and narrowband vowel sound from a sharp transient consonant that begins a word). See [Dau4, Rio] for a more detailed comparison of the STFT and the wavelet transform.

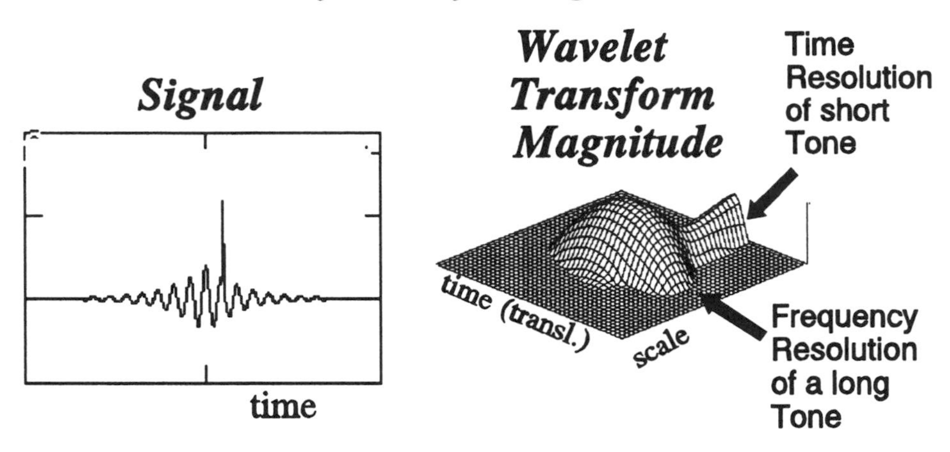

Figure 3.1: Wavelet Transform of Simultaneous Tones

The desirable resolution properties in the previous example are a consequence of the wavelet transform's scaling property that creates a constant fractional bandwidth (constant-Q) filter over multiple octaves. The wavelet "filter" is just the correlation of the signal with a scaled and translated mother wavelet. Essentially, the scale parameter changes the rate of decay on the window function.

As the scale increases, the window duration increases (dilating the wavelet). As the scale parameter decreases the window duration decreases (compressing the wavelet). Just as in Fourier analysis, the translation (time) resolution refines inversely proportional to the duration of the window while the scale resolution refines proportionally to the window duration. Restated, the translation resolution is better as the transform's scale value decreases and the scale resolution improves as the scale parameter increases.

Referring back to the wavelet transform of an impulse with respect to a Morlet mother wavelet, Figure 2.2, the time resolution property can be described. Consider the wavelet domain representation as a set of slices, one slice at each scale value. Then, for the impulse, each slice is just a scaled version of the mother wavelet (the amount of scaling is specified by the scale location of each slice). At small scales near the top of the picture the mother wavelet is highly compressed and achieves very fine time resolution. For larger scales closer to the bottom of the picture the time resolution becomes more coarse and the mother wavelet is more dilated.

When compared to a windowed or short-term Fourier transform the difference is obvious. The window employed in a Fourier transform is a *fixed duration window*. Fourier techniques change the frequency of the oscillating exponential under this window but do not change the duration of the window. The result is a fixed resolution in the time-frequency plane which will not change as the frequency changes. Figure 3.2 shows the time-frequency resolution cells for the short-term Fourier transform. The window size controls the shape of these rectangles but the rectangles will be identical everywhere across the plane; the resolution does not change.

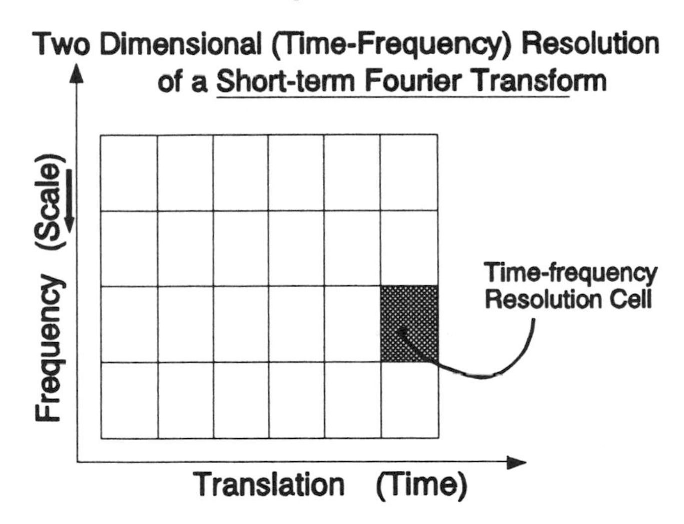

Figure 3.2: Time-scale Resolution of STFT

For the Fourier case, as the frequency increases, the number of cycles of the exponential within the window increases. Whereas, for the wavelet elements, the number of cycles in every element is identical. Thus, for wavelet transforms the "image" of the mother wavelet looks identical at every scale and translation; the "image" is either magnified or defocused to create the scaled version.

The natural feature of the wavelet transform is that it "zooms in" on higher frequency signals and "pans out" on low frequency signals. This feature is appropriate for many naturally occurring signals [Mal1]. However, for some signals a long time duration signal may be desirable at higher frequencies as well. By employing sophisticated mother wavelets with a wide frequency content, simultaneous long time durations and high bandwidths can be achieved.

When dyadic wavelets (scale factors that are integer powers of 2) are employed, the resolution usually follows a pattern similar to that displayed in Figure 2.8. Dyadic wavelets are employed in many multiresolution wavelet transforms and perfect reconstruction quadrature mirror filters (PR-QMFs). When sophisticated and non-dyadic wavelet transforms are employed, the resolution properties of the transform can vary widely. If the mother wavelet is a linear FM signal, then the scale-translation or time-frequency resolution properties will be sloped. See Figure 3.3. The slope is determined by how fast the frequency modulation occurs [Alt, Kel, Spe, Vak, You2]. Compare this sophisticated mother wavelet's (FM) resolution properties in Figure 3.3 to the non-sophisticated Morlet mother wavelet's (Figure 1.1) resolution properties in Figure 2.8. The interpretation of the resolution properties of Figure 3.3 is the same as the interpretation for Figure 2.8 and the discussion of these properties was discussed earlier in Chapter 2.

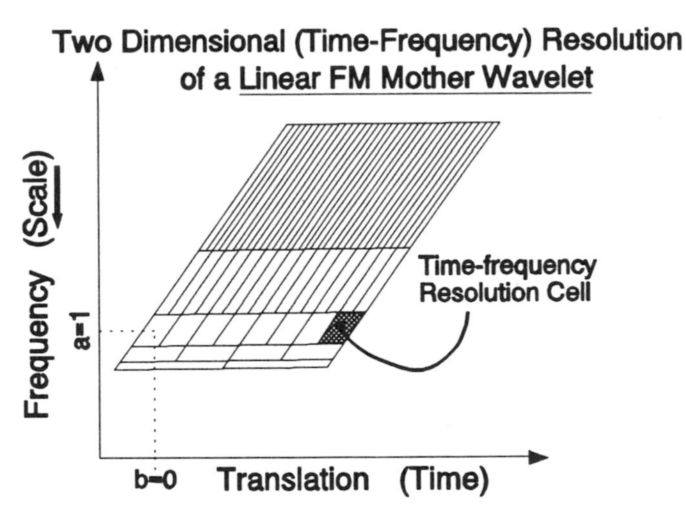

Figure 3.3: Time-scale Resolution of FM Mother Wavelet

Carefully note several features regarding Figure 3.3. *The resolution changes as the scale changes but is fixed across translation for a fixed scale value.* The signal's resolution properties depend upon the scale parameter (for related narrowband processing, the resolution properties across the time-frequency plane do not change with frequency or Doppler shifts). In addition, at the origin, (a=1, b=0), a *fixed* resolution cell exists. This is the resolution in translation and scale of the mother wavelet without any scaling (a wavelet transform of an unscaled and untranslated mother wavelet). Later, this function will be referred to as the auto-ambiguity function.

Figure 3.3 displays an important property of wavelet transforms that is emphasized in this book. The mother wavelet dictates the properties (efficiency, resolution, redundancy, noise immunity, etc.) of the wavelet decomposition. Although all wavelet transforms are constant-Q and multi-scale representations, the mother wavelet controls how the multiple scales are covered and represented. A mother wavelet could have two separate frequencies existing simultaneously. If two simultaneous frequencies exist in a mother wavelet, then, obviously, the time-frequency and time-scale representation will be distinctly different. A single scale value will represent multiple frequencies.

Note that for the dyadic lattice in Figure 2.8, the edges of all of the time windows all align at the center - a straight line at the "middle" translation runs across all scales. This line indicates that, at every scale, the edge of a time window is always occurring at the same point. Due to this edge, some features of the input signal may be lost. A more robust, but non-dyadic, lattice would have scales changing by factors that are less than 2 and create a more staggered edge pattern in the wavelet domain (see Figure 3.3). Thus, the non-dyadic lattice can improve the characterization simply due to its more random or robust window edges. This is important in image coding where the alignment of resolution cell edges is undesirable. By designing the mother wavelet properly, more desirable edge alignment can be achieved.

Multiple Mother Wavelets - Gain and Resolution Properties Only

Later, in Chapter 4, several signals will be decomposed on non-dyadic lattices with respect to different mother wavelets to illustrate the dependence on the mother wavelet and the application of using multiple mother wavelets. Although a specific application will be discussed, the resolution properties can be seen to depend on the mother wavelet as well. When discussing the representation of a function with multiple mother wavelets, the resolution properties of a wavelet transform are critical and the particular mother wavelets that are being employed determine all of the resolution properties (and will control the efficiencies of the representation as well).

Multiple mother wavelets have been considered previously but in a different sense. With the wavelet transforms that have the scale changing as a power of 2 ($a_o = 2$) several researchers [Com, Dau4] have implemented multiple mother wavelet transforms. The motivation of these transforms was to simply *improve the scale resolution but keep the same mother wavelet*. This technique was described earlier and is called *voicing*. A single admissible function is scaled by several scales that are less than 2 to create a set of mother wavelets. This essentially creates a composite wavelet transform with scale values "filled in" by these scaled mother wavelets but still allows the computationally efficient scale factor of 2. This intermediate scaling and application of multiple mother wavelets has been implemented, analyzed and applied [Com, Dau4]. The effective resolution properties in scale are increased and non-octave spacing is obtained. However, the resolution improvement can be sensitive to noise. If the resolution properties of the mother wavelet itself are poor (discussed with ambiguity functions later) then just overlaying a lot of these poor resolution surfaces may not be a good resolution improvement. Voicing is interesting because it uses multiple wavelet transforms but all the wavelet transforms utilize the same mother wavelet (up to a scale factor of less than 2). Efficiency is the justification of this technique and the efficient wavelets often require non-sophisticated mother wavelets. However, voicing with an orthogonal mother wavelet does not yield a composite orthogonal wavelet transform. The "voiced" versions of the mother wavelet will not be orthogonal to the scaled and translated versions of the original (or unvoiced) mother wavelet. The composite wavelet transform becomes nonorthogonal; voicing produces nonorthogonal wavelet transforms even if the original mother wavelet is designed to be orthogonal. The wavelet was designed to be orthogonal on a sparse (scale factor of 2) grid and is then applied on a grid with finer scale resolution; the orthogonality property is lost on this finer grid.

In the more general *multiple mother wavelet* transform, the different mother wavelets are designed to have very different characteristics so as to extract different *features* (portions) of the signal. If the signal has an increasing quadratic frequency modulation (FM) in it and the mother wavelet is a linear down-chirping FM then many wavelet coefficients will be required to represent the signal (an inefficient representation is formed). Whereas, if a quadratically increasing FM is used as the mother wavelet then fewer coefficients can represent the same energy of the signal. As another example, if the application is to detect the presence of the quadratic signal, then, for the down-chirped FM mother wavelet, none of the coefficients in the wavelet domain will have a significant portion of the signal energy (*the gain will be small - no coefficient will have significant value*). Similarly, the resolution will be very poor. Whereas, if the mother wavelet is a quadratically increasing FM, then this wavelet transform will correspond to a wideband matched filter and will provide both gain and a sharp, highly resolved peak. Just as in matched filter theory, or any replica matching scheme, the analyzing function should attempt to match the portions of the signal (extract features that match the replica). When the signal characteristics are unknown, then the set of analyzing functions should be diverse yet provide sufficient gains and resolutions. Thus, when multiple mother wavelets are employed, a diverse set will likely be more desirable than scaled versions of the same mother wavelet (as in voicing). Later in this book an efficient operator for computing multiple wavelet transforms with respect to different mother wavelets is presented. For now the motivation is that the resolution properties of the transform are dependent upon the mother wavelet but more than one mother wavelet may be available.

Consider an example with a relatively long sinusoidal pulse being decomposed by short, medium and long duration Morlet mother wavelets. Consider a tonal pulse with about 50 cycles in it. Now consider a wavelet transform with respect to the Morlet mother wavelet with about 6 significant cycles (i.e., Figure 1.1), another wavelet transform with respect to a 20 cycle long Morlet wavelet and finally a wavelet transform with respect to a 50 cycle mother wavelet.

The transforms are similar but distinctly different. The short duration mother wavelet produces a flat ridge in the wavelet domain while the 50 cycle mother wavelet has a distinct peak. In addition, the peak is significantly higher in the 50 cycle case; more gain has been achieved. The two effects which are being observed is that a longer duration mother wavelet has better resolution and more gain if the input signal is closely matched to it. The gain and resolution are achieved by processing over a longer interval or processing duration. The gain is important in many applications, for example, in data compression the energy should be concentrated in the fewest number of coefficients, in detection, and for general noise immunity where the gain emphasizes the signal energy over the noise energy. Thus, the fixed mother wavelet can be a hinderance similar to the fixed Fourier window. Just as multiple window sizes were desirable for narrowband analysis, multiple mother wavelets can be advantageous for wavelet analysis by providing a variety of cycle lengths and, thus, a variety of resolutions and gains. Obviously, as the mother wavelet changes sizes the density of lattice points in the translation-scale plane can also change.

Consider the example of the last paragraph. Wavelet transforms are touted for their ability to "well" characterize a process over many octaves; however, for this example a long Morlet mother wavelet produced a high gain representation (insensitive to noise) and a short mother wavelet produced very little gain regardless of how much it was scaled. For the Morlet mother wavelet a fixed number of "significant" cycles will exist. The slower the decay and the higher the frequency, the more cycles there are in the mother wavelet. Even for this very simple mother wavelet the wavelet characterization can change drastically by altering the number of significant cycles in the mother wavelet.

Many tradeoffs exist in the multiple mother wavelet situation. A single mother wavelet can have many frequencies simultaneously (a wideband or sophisticated mother wavelet). When a mother wavelet has many frequencies simultaneously, its *translation resolution* (which was previously proportional to the duration of the mother wavelet) now becomes a function of the mother wavelet's bandwidth as well. The translation resolution will improve as the bandwidth increases. When the mother wavelet is a wideband signal with multiple frequency components simultaneously at each time, the duration-bandwidth product of the mother wavelet controls the resolution characteristics rather than just the duration of the mother wavelet.

Mother Wavelet Properties - Relationships to Established Theories

By placing the design of the mother wavelets into a new framework, the choice of mother wavelets becomes the heavily researched area of *signal design* [Vak, Van]. The details of signal design (and thus mother wavelet design) requires the study of *ambiguity functions*. Ambiguity functions are studied later in this chapter. The wideband properties force an extension to classical narrowband signal design and narrowband ambiguity functions which utilizes Fourier transforms and stationary (small variability) assumptions. To examine the resolution property tradeoffs of the wavelet transforms the wideband ambiguity functions of the mother wavelets will be computed.

Besides the resolution tradeoffs, dynamic range tradeoffs exist. The dynamic range of wavelet representations extends well beyond the dynamic range of narrowband or Fourier analysis (due to the wideband signals and the scaling operation). The resolution properties of wavelet transforms remain constant over many octaves [Rio] and thus allow comparison of signal characteristics over many octaves. The resolution properties of Fourier transforms are not constant-Q and thus the resolution properties change across many octaves and do not allow for valid comparisons across wide frequency ranges (dynamic range). Consider a 2000 point FFT of a complex signal with a spectrum that extends from near zero frequency to 20 kHz; this gives a frequency resolution of 10 Hz (20 kHz/2000) across the entire frequency band. A resolution of 10 Hz at 20 kHz is very fine resolution. A resolution of 10 Hz at 10 Hz is very coarse resolution. Demonstrating the resolution tradeoffs, the time (translation) resolution is just the opposite; good at low frequencies and extremely coarse at high frequencies (since the sampling rate must exceed 20 kHz, 2000 points is about 0.1 seconds - 2000 cycles of uncertainty in the 20 kHz region and 1 cycle of uncertainty in the 10 Hz region). The wavelet transform's constant-Q property is achieved with by constantly changing the window size and having a fixed number of cycles in the analyzing "kernel" as discussed in Chapter 1. The wavelet transform provides proportional resolution in each frequency band; as the center frequency increases, so does the bandwidth of the analyzing kernel, the scaled mother wavelet. The translation resolution also changes proportionately, at high frequencies (small scales) the scaled mother wavelet is highly compressed and provides fine time resolution (it "zooms in").

Since mother wavelets can have multiple frequencies (even simultaneously) the time-frequency analysis and bandwidth considerations are not adequate for general mother wavelets. These more sophisticated mother wavelets require a more general approach to characterize their resolution properties; however, since the time-frequency plane analysis is broadly understood and applied, it is used as an analogy and a starting point for the more general wideband ambiguity function analysis.

Throughout the discussion on ambiguity functions the translation-scale resolution will be discussed. The resolution in the translation-scale plane is analogous to the time-frequency plane resolution in the narrowband situation. After the ambiguity function analysis is presented, the resolution properties will be reexamined with a special comparison to the perfect-reconstruction quadrature mirror filters (PR-QMFs) and the related multiresolution mother wavelets.

Signal Analysis - Time-frequency or time-scale Applications

This section briefly discusses wavelet theory's underlying assumptions, advantages and limitations for the application of time-frequency or time-scale signal analysis. Wavelet theory is analogous to short-term Fourier transforms (STFTs), and this analogy is expanded upon in many references to relate wavelet transforms to established, well understood theories [Com, Dau, Fla, Hla, Mal, Rio]. The simultaneous time-frequency or time-scale analysis problem is examined with a wavelet theory emphasis. The resolution characteristics in these two-dimensional planes are emphasized.

Initially assume that the mother wavelet is a windowed tone-like function such as the Morlet mother wavelet in Figure 1.1. These wavelet transforms can zoom-in on high frequency components of a signal; the time resolution becomes very fine for small scale values and correspondingly, the frequency resolution decreases. For analyzing low frequency components the wavelet transforms also pan-out (large scales) to create longer duration windows with poor time resolution but good frequency resolution. Combining the zoom-in and pan-out properties of the wavelet transforms yields constant-Q or constant fractional bandwidth resolution windows.

The product of frequency and time resolution remains nearly constant. The transform's constant resolution processing yields identical gains (constant-Q analysis) over huge operating bands; this is not true in the Fourier transform case. A Fourier transform has the same bandwidth around each center frequency. The fractional bandwidth or Q changes as the center frequency changes and the windowed Fourier transform cannot perform a constant-Q analysis over a wideband.

Due to the extensive research in the time-scale and time-frequency distributions for signal analysis, this application is not pursued further. Please refer to the cited references for details [Hla and references stated in that paper]. However, this research has concentrated on wavelet transforms with tonal-like mother wavelets. With more sophisticated mother wavelets the time-scale interpretation is mandatory; the time-frequency interpretation will not have meaning. But, when sophisticated mother wavelets are employed, the time-scale analysis of a signal is parameterized by the mother wavelet and its properties. For the multiple reasons discussed previously in this chapter relating to the mother wavelet's control over the transform properties, this book emphasizes the ambiguity, uncertainty, or point-spread function viewpoint. This viewpoint allows easier integration of wavelet concepts into existing, and long studied, applications where the ambiguity, uncertainty, and/or point-spread functions are completely integrated.

A Physical Interpretation of Scale-translation Resolution: Wideband Ambiguity Function Conditions

The resolution qualities of the wavelet transform in the scale-translation domain are most easily interpreted as wideband ambiguity function qualities. This provides a strong link to previous research into the narrowband ambiguity function and insight into applications where wavelet transforms may be applicable. The narrowband ambiguity function is closely related to any of the time-frequency techniques (i.e., the Wigner distribution, narrowband ambiguity function, etc. [see Hla for others]). The following wideband conditions also demonstrate that some of the assumptions commonly made in narrowband analysis, such as stationarity, are no longer valid. Details and examples are provided.

Before a detailed discussion of wideband ambiguity functions can be undertaken, both their motivation and their limitations must be presented. These will be presented together because the limitations are most easily stated in terms of a physical example. Wideband and/or nonstationary signals have been extensively researched over the past several decades but have not been incorporated into many systems or products. The problems of the past for processing wideband signals included a huge computational shortfall, some incorrect assumptions, and invalid processing structures or algorithms. Several invalid assumptions are identified to avoid making those mistakes. However, this book does not focus on the problems of the past; the interpretation of wideband processing for future applications is considered. Since wavelet theory is being extensively researched and the interest continues to grow, the wideband processing is formulated with wavelet theory. By using wavelet transforms and associated wavelet operators, the wideband processing can be reformulated and related to wideband ambiguity functions. This book reformulates wideband processing with wavelet theory and also interprets wavelet theory with wideband processing concepts. These interrelationships couple the new wavelet theories with the application-oriented wideband processors. Wideband-nonstationary signals exist in many applications but have been used as the limitation of the narrowband processing rather than part of a model or characterization. By including the wideband-nonstationary features in the modelling (with wavelet theory) some applications will benefit significantly. Applications that can benefit from the wideband or wavelet processing are identified as "wideband" applications.

Wideband Conditions

Applications that can benefit from wideband or wavelet processing satisfy the "wideband conditions." Before discussing the wideband ambiguity function, the conditions that make it a "wideband" ambiguity function are examined. After identifying the wideband conditions, several wideband processors or wideband ambiguity functions will be examined. Each different wideband ambiguity processor involves a unique set of assumptions or conditions. To accurately discuss the assumptions required for various ambiguity function forms, the assumptions must be defined and developed. First, the conditions on the signals are briefly stated in terms of fractional bandwidths and then more accurately with duration-bandwidth products. In the following sections a detailed analysis of the wideband conditions for *multiple octave signals* is provided and then related to the various computational forms for wideband ambiguity function processing. This section provides the motivation for wideband processing without wandering too far away from its relationship to wavelet transforms. When is a signal or a system considered wideband? For many applications the wideband condition depends on both the signal and the system together.

Wideband Signal Conditions

Typically, the extent of the signal bandwidth compared to its center frequency, the fractional bandwidth (BW/f_c), determines if a signal is wideband (the fractional bandwidth is a signal related measure). The fractional bandwidth (sometimes referred to as the signal "Q") determines when a *phase and amplitude envelope* "riding" on a complex exponential at a center frequency is an acceptable model. A **modulated signal** refers to a sinusoidal or complex exponential signal at a particular carrier (center) frequency, ω_c, that has been encoded (modulated) with information; typically the carrier frequency is much higher than the bandwidth of the modulating signal. The modulation or encoding is usually performed by modifying the sinusoid's amplitude, phase or both over time [Van]. This phase and/or amplitude modification is the modulating phase and amplitude **envelopes**, $\phi(t)$ and $a(t)$, respectively. In terms of these modulating envelopes the modulated signal can be expressed as:

$$f(t) = a(t)\, e^{j\phi(t)}\, e^{j\omega_c t} = a(t)\, e^{j(\omega_c t + \phi(t))} \qquad (3.1)$$

Different modulation types, such as frequency modulation (FM), phase modulation (PM), and amplitude modulation, directly encode the modulating information into the envelopes. Other modulation types simultaneously modify both the amplitude and phase envelopes.

When the envelopes are modified at a rate (highest frequency in the modulating signal) that is significant compared to the center frequency, then the center frequency should no longer be separated from the envelopes - the signal is considered *wideband.* A significant modulation rate would be greater than about a tenth of the center frequency. If the phase variation in the envelope occurs at a rate that is near the phase variation occurring in the carrier frequency term, then the envelope and carrier phase variations must be considered jointly. This fact will be more rigorously developed after the center frequency and bandwidth are more rigorously defined. The primary result is that, for wideband signals, the carrier frequency term cannot be separated from its envelope without introducing unacceptable distortion (if the distortion is acceptable, then the signal is considered as narrowband).

Wideband Signals and the Analytic Signal Model

The justification for this fractional bandwidth wideband condition is formulated with the analytic signal representation [Kel, Swi, Van1] which uses the signal envelopes to represent the signal and, effectively, discards the exponential at the center frequency. In many applications, to reduce the number of samples and the sampling rate researchers employ this *analytic* signal model to mix the signals to their baseband equivalents (envelopes-only representation). The analytic signal envelopes are then processed by the appropriate transforms and other operations (including wavelet transforms). This envelope processing is *invalid* for wideband signals. The invalidity of this basebanding processor is subsequently justified.

The analytic signal is often formulated with a Hilbert transform [Bos, Van1]. The **analytic signal** is formed from a real signal, $f_{real}(t) = a(t)\cos[\omega_c t + \phi(t)]$, by adding the real signal to the Hilbert transform of the same real signal ($f(t) = f_{real}(t) + j \cdot HT[f_{real}(t)]$). For this function to have the form as given in equation (3.1), it requires *the condition of no spectral overlap*. If a signal is modelled as a product of a high frequency signal (chosen as the tone or complex exponential at the carrier frequency, $e^{j\omega_c t}$) and a low frequency signal (the envelope signal, $a(t)\, e^{j\phi(t)}$):

$$f(t) = lf(t) \cdot hf(t), \text{ or specifically} \\ f(t) = a(t)\, e^{j\phi(t)}\, e^{j\omega_c t} \tag{3.2}$$

then only if (no spectral overlap between the low frequency and high frequency signal portions):

$$LF(\omega) \cdot HF(\omega) = 0, \quad \text{for all } \omega \tag{3.3}$$

will the Hilbert transform of the product of a high frequency signal, $e^{j\omega_c t}$, and a low frequency signal (the envelope signal, $a(t)\, e^{j\phi(t)}$) be the low frequency signal times *the Hilbert transform of the high frequency signal*:

$$HT[f(t)] = lf(t) \cdot HT[hf(t)] \tag{3.4}$$

where HT denotes the Hilbert transform [Van]. For the modulated envelope this becomes:

$$\begin{aligned} HT[f(t)] &= lf(t)\, HT[hf(t)] = a(t)\, e^{j\phi(t)}\, HT\left[e^{j\omega_c t}\right] \\ &= a(t)\, e^{j\phi(t)}\, e^{j(\omega_c t - \frac{\pi}{2})} = e^{-j\frac{\pi}{2}}\, a(t)\, e^{j\phi(t)}\, e^{j\omega_c t} \\ &= e^{-j\frac{\pi}{2}}\, f(t) \end{aligned} \tag{3.5}$$

Note that the Hilbert transform of an exponential is just a phase shift [Bos1, Van] and that the center frequency was assumed to be positive.

When the real signal is added to its Hilbert transform, the phase shifts interact so as to annihilate the frequency components less than zero and the analytic signal is created. All of these operations are under the assumption that no spectral overlap exists. For this model, that implies that the center frequency is larger than *any* frequency content of the modulating signal.

The analytic signal model is invalid for some wideband signals. The Hilbert transform of a product of a tone (at the center frequency) and a lowpass (envelope) signal does *not* necessarily equal the product of that tone and the Hilbert transform of the lowpass signal. This condition is required for a valid analytic signal representation of a bandpass signal. The required condition is that the center frequency of the tone be at a frequency that is higher than any of the frequencies of the lowpass signal with "significant" energy (possibly as low as 20-30 dB down). The center frequency must be greater than this "extended bandwidth."

When the center frequency is less than the "extended bandwidth" then the analytic signal model will be invalid. The signal cannot be represented as a product of a high pass and a low pass signal. If the operations to form an analytic signal are still performed, then the resulting signal does not have the correct phase and amplitude envelopes (these envelopes will not match the envelopes of the original real signal). It can be shown that *a 180 degree phase shift is applied to high frequency portions of the signal's spectrum beyond the center (or high) frequency.* The effect of multiplying the high frequency components by a minus one can be significant - much of the signal information is contained in the phase (including all of the time localization properties of the Fourier representation). All time samples are affected by changing these frequency components. If the analytic signal model is employed with wideband signals, the analytic signals cannot be formed by adding the real signal and its Hilbert transform.

Effective rms Time-bandwidth Product

For a more rigorous statement of the wideband conditions for signals, the time-bandwidth product is considered. For most wideband signals, the signals are considered *wideband* if their fractional bandwidth, the bandwidth divided by the center frequency, is greater than approximately 0.1 (or 10%) or $\left[\frac{(BW)}{\omega_c}\right] > 0.1$ when $\omega_c \neq 0$. If the center frequency is zero, the definition does not apply and the signal is considered wideband. For a signal, $f(t)$, the spectrum is defined as the Fourier transform of $f(t)$, $F(\omega)$. The center frequency is defined as the *spectral centroid*:

$$\omega_c = \left[\frac{\int_L^{\infty} \omega |F(\omega)|^2 d\omega}{2\pi \int_L^{\infty} |F(\omega)|^2 d\omega}\right] \quad (3.6)$$

where $L=0$ for real signals and $L=-\infty$ for complex or analytic signals. The bandwidth is defined as the *root-mean-square (rms) bandwidth*:

$$BW = \sqrt{\frac{\int_{-\infty}^{\infty} \omega^2 |F(\omega-\omega_c)|^2 d\omega}{(2\pi)^2 \int_{-\infty}^{\infty} |F(\omega-\omega_c)|^2 d\omega}} \quad (3.7)$$

but for real signals only the positive frequency portion of the original spectrum is included in this integration. Note that signals with energy at low frequencies (lowpass signals) are always considered wideband signals; this is due to the rapid changes in scale occurring (many octaves) around zero frequency.

For the analytic signal discussion in the previous section, the center frequency definition should be the rms-center frequency defined in equation (3.6). But the rms-bandwidth may *not* be an acceptable definition for the bandwidth of the modulating signal due to its definition as the 3 dB bandwidth. Bandwidths are somewhat arbitrary until specific attenuations at the extremes of the band are specified. The energy of the lowpass modulating signal must be significantly attenuated at the center frequency; 3 dB is not sufficient. An acceptable distortion level will determine the extent of the required attenuation (about 20-30 dB at least) and this attenuation will lead to a larger bandwidth.

For any signals that are truly *multiple octave signals, the analytic signal will be an invalid representation.* Voice is an example of a multiple octave signal, with frequency components less than 100 Hz and greater than 6.4 kHz (6 octaves). An analytic signal generated from a voice signal will not accurately model the voice signal - some frequency components will be phase shifted and others will not (as demonstrated in the previous Hilbert transform analysis).

What is the affect of this distortion? If the signals are multi-octave signals and the analytic signal form is used, *the demodulated lowpass envelope signal will be different than the original lowpass modulating signal* (as discussed above). Thus, for active sensing, where the transmitted modulating signal was known, the replicas that are formed from this original transmitted lowpass signal will *not* match the received signal envelope. The mismatch between the replicas and the received envelopes will, in most cases, cause the maximum correlation gain or resolution performance to be significantly degraded (matched filter processing performance degrades). In conclusion, the analytic signal model and associated envelope processing is valid only when the lowpass signal's "extended bandwidth" is less than the bandpass signal's center or carrier frequency.

Returning to general signal properties, besides a signal's fractional bandwidth, the signal's effective time duration is also important. For a signal, $f(t)$, the center time, T, is defined as:

$$T_c = \left[\frac{\int_{-\infty}^{\infty} t |f(t)|^2 dt}{\int_{-\infty}^{\infty} |f(t)|^2 dt} \right] \tag{3.8}$$

The effective time duration is defined as the *root-mean-square (rms) duration*:

$$T = \sqrt{\frac{\int_{-\infty}^{\infty} t^2 |f(t-T_c)|^2 dt}{\int_{-\infty}^{\infty} |f(t-T_c)|^2 dt}} \tag{3.9}$$

The center time can effectively center the position of the signal at zero time. The effective time duration measures the extent of the signal's time duration.

The rms bandwidth and time duration can be combined to form an effective, rms time-bandwidth product, $T \cdot BW$. When this time-bandwidth product is considerably greater than 1, then the signals will be considered *wideband*; Vakman [Vak] calls these signals **sophisticated signals.** These signals can be deterministic or stochastic; if the signals are stochastic, then expected value of f(t) replaces f(t) in equations (3.8) and (3.9) and the power spectral density of the f(t) replaces $|F(\omega)|^2$ in equations (3.6) and (3.7).

Wideband Systems and Signals

For *systems* involving motion and travelling waves the wideband conditions are related to the ratio of the maximum motion in the system compared to the speed of travel of the wavefront in the system, max_speed/wavefront_speed. However, for this book the wideband condition will simply be formulated as the *narrowband condition being invalid*; combining the signal and system conditions, the wideband condition becomes the *invalidity* of the following narrowband condition:

$$\frac{2\ max\_speed}{wavefront\_speed} \ll \frac{1}{BW\,(signal\_duration)} \qquad \textbf{(3.10)}$$

where BW is the root-mean-square bandwidth of the signal, max_speed is the highest relative speed between any elements in the system, wavefront_speed is the rate at which the wavefront advances (i.e., speed of sound in a medium, speed of light), and signal_duration is the time duration of the signal. Note that the right side of this equation is simply the reciprocal of the time-bandwidth product. The left side of the equation is the fastest, normalized speed that the wavefront can encounter in the system (environment).

What happens to the narrowband processor when the condition in equation (3.10) is invalid? The gain of the processor decreases. Since the processors are correlation processors then the gain of the processor is proportional to the valid processing interval. As the interval increases, the gain becomes higher. If the model is invalid over part of the processing interval, then for the invalid interval, no signal gain is achieved; however, more noise energy is introduced in that interval and, thus, the processor gain decreases. The magnitude of the ambiguity surface may not have a significant peak [Van]. The narrowband replica that is created by the narrowband processing only approximates the "higher order" time warping (which, in general, is not just a linear time scale). For the reflection example, the true reflection action is to time map the signal with some time scale and possibly higher order acceleration terms (some nonlinear function in general) [Cham, Kel]. When the signals are narrowband, frequency shifting can approximate the time scaling operation, but frequency shifting does not approximate time scaling for wideband signals. Thus, for wideband signals the frequency shifted replicas will not match the reflected signal and the gain of the processor approaches zero.

Since scale is the parameter of interest, these wideband conditions can be restated in terms of scale limits as well. The reformulated conditions interpret the ratio of max_speed/wavefront_speed as the maximum possible scale that can act on the signal. The reciprocal of the time-bandwidth product of the signal represents the scale resolution of the signal. The narrowband condition requires that the maximum possible scale due to the motion be much less than the scale resolution of the signal. Visualizing with physical parameters instead, the range resolution of the signal (1 over the time-bandwidth product) must cover the maximum region over which the system can move (max_speed times the signal_duration). See Figure 3.4. The system/signal combination is considered wideband when the signal's scale resolution is comparable to or less than the possible scale due to system motion. With physical parameters the system/signal combination is wideband when the signal's resolution is comparable to or finer than the possible range change in the system (max_speed times signal_duration). See Figure 3.5. Using these conditions and the bounds on the motion in the environment, a signal designer can determine limits for the signal's bandwidth and duration to maintain the narrowband requirements. As the motion increases the time-bandwidth product of a valid narrowband signal shrinks (causing the signal to have less gain and poorer resolution characteristics). For some configurations, the signals may not be alterable. For these cases the wideband condition must be checked to determine if narrowband or wideband processing is applicable. It will be important later in the book to understand that the signal's duration imposes the region of validity for the narrowband model.

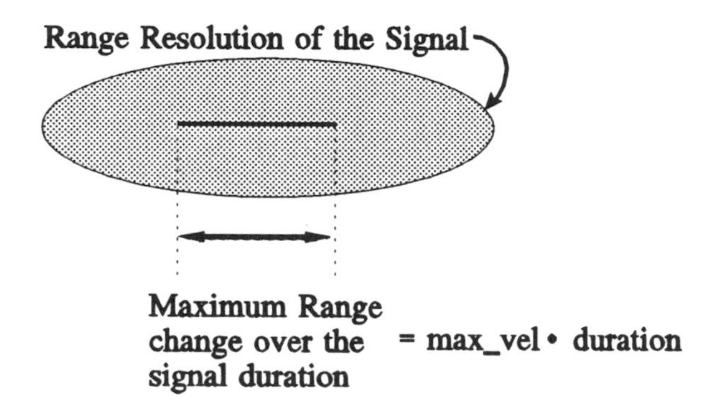

Figure 3.4: Narrowband Condition in Space

Figure 3.6 demonstrates that a single point scatterer can appear to be at different ranges depending upon the chosen time reference. A single point scatterer should be represented by only one point (a single range) in the wideband characterization (the range map for characterizing an environment). Each different time reference leads to a different location of the scatterer. This example is presented for motivation; it is not completely described or characterized (that analysis requires the wideband/wavelet system modelling of Chapter 5). But this example should help differentiate the wideband/wavelet processing from the narrowband processing.

Wideband Signal/System Condition

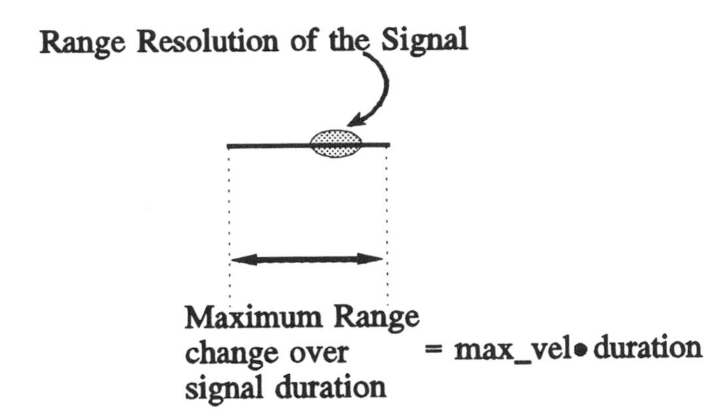

Figure 3.5: Wideband Condition in Space

Wideband Signal/System Reference

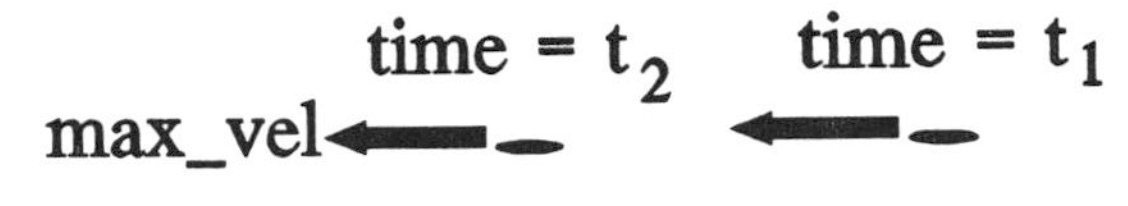

location (range) of the scatterer changes over the signal duration (different at the beginning and the end of the signal)

Processing time, T_p, is $(t_2 - t_1)$, where t_1 is the beginning of the signal and t_2 is the end of the signal duration

The characterization of the environment should only have 1 point for 1 point scatterer - the environment is characterized at only one snap-shot in time (not through the entire signal duration)

Figure 3.6: Wideband Time Referencing

For the previous three figures the range (or translation) resolution of the signal compared to the possible motion in the system sets the limits on the required processing models (narrowband or wideband). If, instead, the maximum valid process processing interval, T_p, is considered, then the limits on the maximum valid processing duration can be established and compared for both the narrowband and wideband models. The maximum valid processing duration is the interval over which the model is valid (the validity is established by meeting the conditions shown in Figure 3.4 for the narrowband model). It will simply be stated that the wideband model is limited only by acceleration and this derivation is deferred until Chapter 5. The limitations on the narrowband and wideband models are, respectively:

$$T_{p\text{-}NB} < \frac{wavefront\ speed}{2 \cdot bandwidth \cdot velocity} = \frac{c}{2 \cdot BW \cdot vel} \qquad (3.11)$$

$$T_{p\text{-}WB} < \sqrt{\frac{wavefront\ speed}{2 \cdot bandwidth \cdot acceleration}} = \sqrt{\frac{c}{2 \cdot BW \cdot accel}} \qquad (3.12)$$

Note that the wideband model has an infinite valid processing interval when the system only has constant velocities (zero acceleration) while the narrowband model will be valid only for some finite interval dependent upon the speeds in the system (or environment) and the signal's bandwidth. Also observe the similarities of the two equations; if acceleration simply replaces velocity in equation (3.11), then the two equations become identical except for the square root. The wideband model can account for the linear time variation induced by the velocities but cannot account for the higher order time variations such as acceleration. The wideband processing's extension of the valid processing interval leads to higher gains (more time, more energy, etc.) and better resolution. The extension of the valid processing interval can be interpreted with the previous three figures. For Figure 3.4, the system or environment is assumed to be motionless over the duration of the signal - the poor signal resolution enforces this motionless requirement. For Figure 3.5, the long processing interval and fine signal resolution allows the motion in the system to be resolved over the duration of the signal. This system or environment is time-varying (nonstationary if it is stochastic) over this processing interval; the scatterer is at a different location or range when the signal's front edge reflects versus when the signal's trailing edge reflects. The system cannot be assumed to be motionless (narrowband or, if stochastic, stationary) when this processing interval is used. The high resolution can only be employed relative to one time instant (snap-shot or time reference). Several time references can be used; the beginning of the signal, the end of the signal, or the middle of the signal are typically employed.

Summarizing, the narrowband processing can be used to model a system or environment but its processing duration is limited by the signal's bandwidth and the motion (or changes) in the system or environment. The wideband modelling is not limited by velocities and its processing duration is limited only by accelerations. The wideband model "images" the system or environment at only one specific time instant (snap-shot). However, a long processing interval can be employed to estimate the "image" or model at that instant of time. The advantage of the larger processing interval is more gain (energy), noise immunity, better resolution, etc. Any time-varying or nonstationary model is necessarily referenced to some time reference. Later in Chapter 5 the details of the time referencing are analyzed. To justify the wideband model, consider the following example.

Consider a motivational example. Assume that a person is driving a motorcycle directly at you at about 25 mph and screaming something to you (see Figure 1.7). Due to the bandwidth of the voice signal being several kHz and the speed of sound in air being only about 730 mph, the narrowband valid processing interval is on the order of several milliseconds at most (from equation (3.11)). If the signal is processed for longer than the narrowband valid processing interval, then the model may be invalid and lead to losses and inefficiencies. For the longer processing intervals the narrowband frequency shifting model is not valid over large bandwidths. The problem is that the closing velocity of the motorcycle causes a time scaling or compression of the voice signal. The time scaling will shift different frequency components by different amounts (the frequency shift is proportional to the center frequency). Thus, the voice signal that you hear cannot be well modelled by the narrowband model, so that the short processing intervals can lead to poor models.

Now consider a wideband model. Consider a wideband system that receives sound and time scales the sound signal. Assume a little "black box" (wideband system) exists that converts the received signal (box input) into an output that is a time scaled version of the received signal. If this little box dilates the signal to exactly offset the compression introduced by the motorcycle's motion, then the output from the box sounds just like the rider's speech in the rider's reference frame (ignoring the effects of a 25 mph wind in the rider's face). Thus, up until the motorcycle strikes you (or passes by you), the little black box converts the received warped speech into comprehendible speech. The little box acts as an inverse system to account for the motion of the motorcycle (the physical system).

Consider a different problem, called echo-location. The goal of the echo-location is to estimate the position (or range) of the motorcycle. Due to the significant motion (relative to the speed of sound in air) and the large bandwidth, the wideband model is employed. For echo-location a signal is transmitted and the echo is processed to locate the scatterer (the motorcycle). To simplify the explanation, assume that the transmitted signal is the same scream that was discussed in the previous paragraph but emanating from the motionless person in Figure 1.7. Now the reflected or echo signal from the motorcycle is the same compressed signal received in the previous discussion and, thus, a little "time-scaling" box can be built to output the "unscaled" or original signal. Since the transmitted or original signal is known to the motionless person, then the output of the box and the transmitted signal can be correlated over a long period of time (valid processing interval) and produce a high correlation or high gain. If noise is present (wind noise at the ear or microphone) then the gain can be used for noise immunity and robustness in the echo-location. In addition, the longer valid processing interval produces better scale resolution (detailed subsequently).

The correlation processing will hypothesize across both scale (velocity) and delay (range). A peak of this correlation surface will indicate that the signals are the same. The delay and scale at which this peak occurs corresponds to the range and velocity of the closing motorcycle. However, depending upon the structure of the correlation processor, the delay can be referenced to different points in time and, thus, correspond to different motorcycle ranges. The reference frame discussion is also deferred to Chapter 5. By accounting for the time variation or nonstationarity, a long processing interval can be used to achieve high gain and resolution, but processing over the entire interval yields an estimate for the range at one specific time reference only.

Notice that the motorcycle was assumed to directly approach the standing person (approach with a radial velocity only). For constant velocities in the system, the wideband valid processing interval is unlimited. As soon as any accelerations occur, the wideband model begins to degrade (analogous to the narrowband model degrading due to velocities in the system). Although the limitations of the wavelet model are not detailed until Chapter 5, it should be noted that accelerations limit the wideband or wavelet processing just as velocities limited the narrowband model. When an acceleration occurs, the time scaling will change over the duration of the signal. By limiting the duration of the signal, the acceleration can be made negligible (this is the wideband limit in equation (3.12)); however, by limiting the duration of the signal even further, the narrowband model could be used as well. Although the wavelet or wideband model extends the valid processing interval, the conclusion is that wavelet or wideband processing accounts for only the first order or linear time warping and the higher order time warpings will limit the valid processing duration of the wideband or wavelet processing. Details of these limitations are further discussed in Chapter 5.

Active and Passive Sensing

Both active and passive sensing schemes can employ wideband or wavelet processing. Figure 3.7 shows the active sensing system with the transmitter and receiver being collocated and the transmitted signal assumed to be wideband. Under the assumption that the reflector's velocity and the transmitted signal's parameters cause the narrowband condition to be invalid (equation (3.10)), then wideband processing will be required. To a first order approximation (and many assumptions regarding the transmission environment), the reflected signal will be a time scaled and delayed replica of the transmitted signal. The received signal is then compared to the transmitted signal; if this comparison is a correlation, then a wideband cross correlation is formed between the received and transmitted signals. This correlation creates a wideband cross ambiguity function (later this is shown to be a wavelet transform as well).

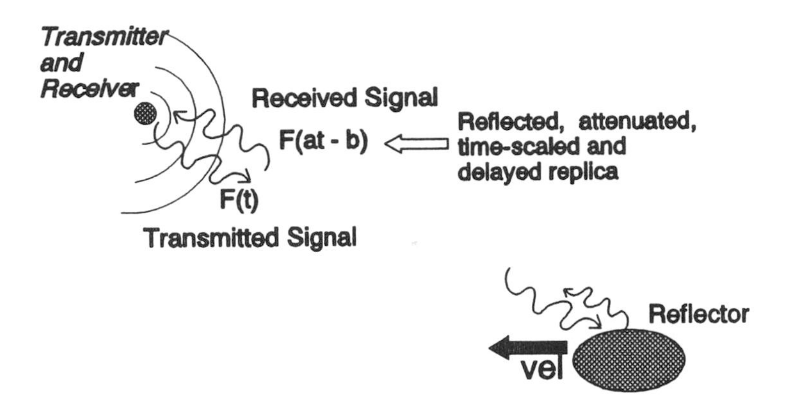

Figure 3.7: Active Wideband Sensing

For the passive system no transmitted signal exists and only two or more received signals are available. See Figure 3.8. If one of the received signals is chosen as the reference signal then it can be scaled and translated and compared to (or correlated with) the other received signal(s). Again, the comparison forms a wideband cross ambiguity function.

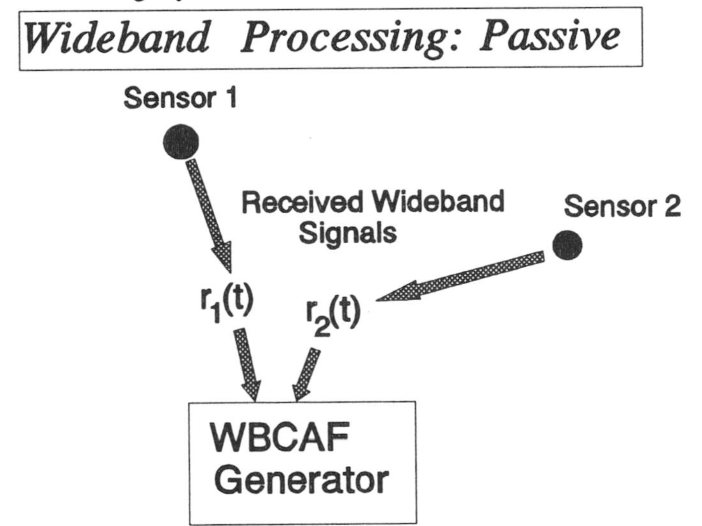

Figure 3.8: Passive Sensor Wideband Processing

For either passive or active processing the wavelet or wideband processor is nearly identical. Only the signal employed as the "replica" or mother wavelet is different. See Figure 3.9. For the active case the mother wavelet is the transmitted signal, while for the passive sensing the mother wavelet is one of the received signals.

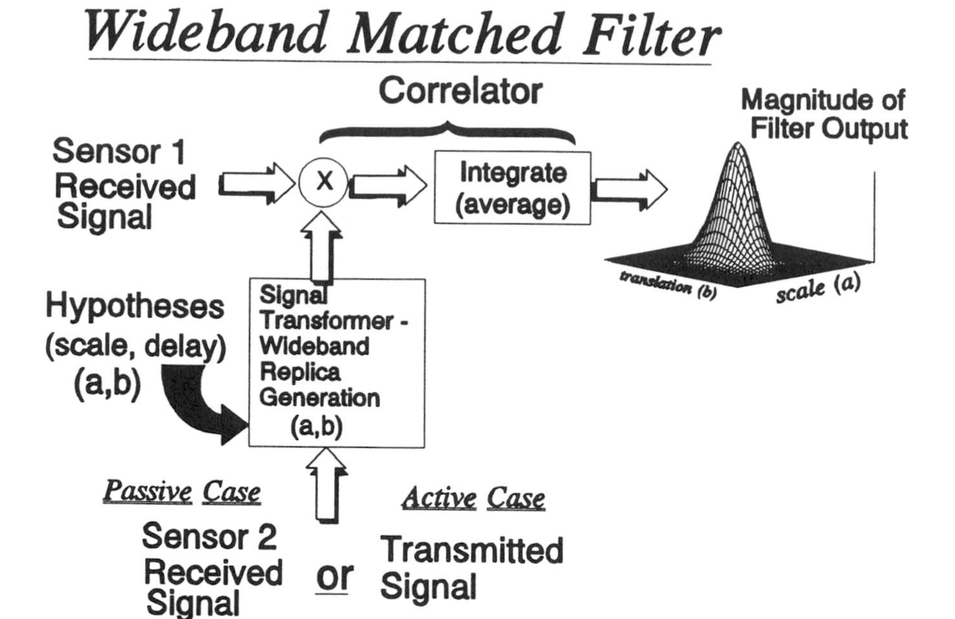

Figure 3.9: Wideband Matched Filter

These passive and active sensing models are extremely general and can be applied in many fields. However, the areas where it may have noticeable impact on performance are applications that have significant motion or time-space-variation.

Although some readers may respond to the wideband conditions by thinking, so what? The impact of this wideband condition is significant. Besides the narrowband assumptions being invalid, the assumption of a stationary (both motionless and wide-sense stationary) environment are also eliminated. Time and/or space-varying models are required. Why is this drastic step required? Reexamine Figure 3.5. When the signal resolution is finer than the possible motion in the system, then it cannot be assumed or expected that an observation either earlier or later in time will be similar, yet that is the stationarity assumption. Note that this is not the distributed scatterer case; a point scatter with significant motion can satisfy the wideband condition. The maximum travel range is the distance that even a point scatterer may have traveled over the signal duration. This condition is valid for point scatterers and states that the scatterer's position will, most likely, be different during the next observation interval, *relative to the positional resolution of the signal.*

The wideband condition is often related to nonstationary signals and systems for this reason; the statistical properties change over time. Sometimes the statistical interpretations are difficult when the signals are deterministic, so this condition is referred to as the wideband condition, in general. The wideband condition will also invalidate the classical scattering function models [Van, You2, Zio] due to the model's wide-sense stationarity assumption (more on these assumptions later). Of course, linear time invariant system models are invalid and convolution is no longer the correct operation. Later, in Chapter 5 a more general operator is derived and justified that can handle both the time and/or space-varying conditions and the nonstationarity (it is general, so it also handles the stationary or space-time-invariant cases as well).

Ambiguity Functions

The wavelet transform can be formulated as a wideband cross ambiguity function (processor). The signals to be "cross" processed are the mother wavelet and the received signal. In passive sensing configurations, two received signals may need to be cross processed but time scaling a received signal is an extremely computationally intensive task. A later section in this chapter derives a wavelet domain technique that efficiently forms the wideband cross ambiguity function between two received signals. The wavelet domain reformulation has the exact same structure as the FFT reformulation of the narrowband cross ambiguity function. Details are presented in the following sections.

This book utilizes ambiguity functions for determining the resolution properties of wavelet transforms. The interpretation of the wavelet transform as a replica correlator processor (mother wavelet is the replica) leads to the ambiguity function controlling its resolution properties. But the ambiguity functions are wideband ambiguity functions in scale and translation instead of the narrowband frequency and translation parameters. Wideband ambiguity functions cannot be computed as easily as narrowband ambiguity functions and, if the assumptions are not satisfied, invalid wideband ambiguity processors can be formulated.

Ambiguity Function Forms, Assumptions, Computations and Tradeoffs

Ambiguity functions have many forms including combinations of wideband/narrowband, auto/cross and can be formulated with Fourier, FFT and wavelet transforms. However, every ambiguity function is essentially a correlation of an "unmodified" signal with a second "modified" signal. The output of the ambiguity function is a surface of correlation values; one correlation for each "modification" (e.g., time shifting and time scaling or time shifting and frequency shifting). Each form of the ambiguity function and each signal model requires a variety of assumptions; when the appropriate assumptions are invalid, the corresponding processor is invalid. The wideband ambiguity function is sometimes incorrectly formulated and implemented due to invalid assumptions and models. Appendix 3-A identifies the forms and assumptions for each ambiguity function processor and its associated advantages, efficiencies, and tradeoffs when compared to other ambiguity function processors. Finally, ambiguity function implementation techniques with wavelet, Fourier and FFT transforms are discussed, highlighting possible invalid assumptions. These ambiguity function implementations are closely related to wavelet transforms, as the next section exemplifies, but they are not included as an entire section here due to the specifics of implementation and computation. The sections of this book concentrate on the theoretical aspects and the integrated applications of wavelets, but avoid most of the detailed implementation considerations. So please refer to Appendix 3-A for detailed considerations of ambiguity function forms and their implementation.

Ambiguity function analysis is used in many fields, ranging from signal design to positioning of stars. Ambiguity function analysis is utilized to determine the resolution and/or gain of many systems and also to characterize the effects of systems or channels (scattering functions). Applications such as biomedical, speech, geophysical, oceanographic and other acoustical sensing systems require the signal processing of wideband, multi-octave signals. Often the environments in which the signals are sensed or the sensor configurations themselves are nonstationary or in some type of motion. By coherently processing these signals over long observation intervals the gain and/or resolution of the overall processing or imaging system may be significantly improved; however, the long observation intervals again lead to nonstationarities or time-varying spectra in the observed signals. Analogous conditions hold for the spatial parameters and the spatially-varying wavenumber spectra but the spatial dimensions will not be pursued.

Ambiguity functions represent the similarities between one signal and many modified versions of another signal. When the two signals being compared are the same, an auto-ambiguity function is formed, when the signals are different, a cross ambiguity function is formed. When the signal modifications are time scaling and delay (translation), then the wideband ambiguity function is formed. When the modifications are time and frequency shifting (translation in both time and frequency), then the narrowband ambiguity function is formed. If the signal and/or environmental characteristics require wideband representations, only a wideband ambiguity function is valid. Under a more restrictive set of assumptions, narrowband ambiguity functions are valid. Under another further set of assumptions (regarding both the signals and the noise), certain forms of the ambiguity function are optimal position and velocity (state) estimators [Van, Zio].

For wideband ambiguity functions, time scaled signal versions are formed. When a wideband scaled replica is created, the bandwidth of the replica is different than the bandwidth of the original signal. If the signal is scaled by a factor of 2 then the bandwidth is also changed by a factor of 2 as well. Thus, when large scale values are to be used in creating the wideband ambiguity function, the bandwidth and sampling of the original signal must account for these changes. For discrete signals aliasing effects must be avoided. Obviously, the time duration of the signal replicas is similarly affected. The rms time-bandwidth product remains constant.

Wavelet Transforms and Ambiguity Functions

Wavelet theory has extended fields other than just transform representations. The theory surrounding ambiguity functions has been significantly improved. Wavelet transforms are now included in the large class of time-frequency representations such as the short term Fourier transform (STFT), the Wigner distribution (and associated weightings), and the narrowband cross ambiguity function. The previous section and reference [Hla] address many of the relationships and properties of these distributions and their applications. Wavelet transforms are essentially defined as wideband cross ambiguity functions and this similarity and effect on the ambiguity theory is expanded upon. For those understanding group theory, the wideband cross ambiguity function (WBCAF) is simply the left-regular wavelet transform [Hei], but this viewpoint is not expanded upon any further due to the specialized mathematical requirements. The justification for examining wideband ambiguity functions was previously mentioned.

Ambiguity functions represent the similarity between one signal and many modified versions (modified *replicas*) of another signal. The "similarity" is measured by a correlation. In passive sensor processing both of the signals may be unknown and received, but portions of the signals may originate from the same source or scatterer. For active sensor processing one signal may be the transmitted signal while the other signal is the received signal. When the two signals are the same, an auto-ambiguity function is formed, when the signals are different, a cross ambiguity function is formed.

Wideband Cross Ambiguity Functions

The **wideband cross ambiguity function (WBCAF)** is defined for two signals, $r_1(t)$ and $r_2(t)$. Assuming $r_1(t)$ is an admissible function, the WBCAF maps two finite energy one-dimensional signals into a finite energy, two-dimensional plane ($\{L^2(\mathbf{R}) \times L^2(\mathbf{R})\} \to L^2(\mathbf{R}\backslash\{0\} \times \mathbf{R})$) as:

$$WBCAF[r_1, r_2](s,\tau) = \sqrt{s}\int r_2(t)\, r_1^*(st-\tau)\, dt \tag{3.13}$$

The WBCAF parameter "s" is the differential scale parameter and "τ" is the differential delay parameter. "Differential" refers to the differences between the scale and delay parameters between the two signals being processed and not to the characteristics of either signal alone.

This WBCAF is referred to as the "scale-then-delay" form because the first signal, $r_1(t)$, is scaled and then delayed (or advanced) before being correlated with the second signal. If parentheses are inserted around the $t-\tau$ term so that the scale acts on both parameters, then that form is called the "delay-then-scale" form. These forms are compared in Chapter 6.

Physically the WBCAF represents the correlation between one signal and scaled and translated versions of a second signal. In an active sensing problem the one signal will be known since it is transmitted and the other signal will be a reflected, refracted, and/or reradiated version(s) of the transmitted signal. When the two are correlated, a single correlation coefficient is output at each scale and translation value; these values will later be interpreted as a representation of the "scattering process" and details of this interpretation are deferred until Chapter 6. A second physical interpretation is that the two signals are received at two passive sensors. Now only the portions of the signal common to both received signals will be identified by the correlation process. This passive processing can be used to passively position emitting or reflecting sources (image an environment). A final example of the "cross wavelet" or WBCAF processing is the identification of a system or channel model (creating a time varying characterization of the system or channel) - this would be applicable in a speech recognition system. Again the details of this application are deferred until Chapters 5 and 6. First the theory of WBCAFs must be understood.

One of the first derivations of the WBCAF was for the active case of a transmitter and receiving sensor being collocated [Kel]. The transmitted and received signals are the two signals for this type of processing, that was later generalized [Alt, Zio] for signal design in stationary environments. For these derivations and processors, many assumptions must be made. Instead of making all of these assumptions and deriving the WBCAF, this book simply refers to the references for those assumptions and derivations. As mentioned previously, higher order acceleration terms (a nonlinear function in general) are required to exactly model the reflection process. So instead of making any assumptions regarding a particular problem, the WBCAF is just defined as an operator - without any physical justification. Derivations of the WBCAF for the specific problem of reflection (under specific sets of assumptions) can be found in the references [Alt, Kel, Van] although [Van1] immediately jumps to the narrowband case. This book keeps the WBCAF as a general operator and simply evaluates this operator under different sets of assumptions to produce results similar to previous results. The narrowband case is simply the addition of another assumption and thus it is just another evaluation of the WBCAF.

Several other "forms" of the WBCAF have been defined, but for this book the only definition of the WBCAF is equation (3.13). Alternative forms include the wavelet transform and the "delay-then-scale" form that places parentheses around the $t-\tau$ argument. The WBCAF form in equation (3.13) was chosen for several reasons. This WBCAF form is usually easier to implement than the wavelet transform because the signals can more efficiently be scaled and then translated instead of translated and then scaled. When processing blocks of data rather than the entire data stream, a "bulk" or block delay is initially performed and then the WBCAF acts only on the data within the block (this processing is done almost exclusively in passive sensor processing). Since the block delay occurs first, the majority of the delay term will act as a "delay-then-scale" form but to be complete, the block processing with the WBCAF is a "delay-then-scale-then-delay" form with the first (block) delay usually being much larger than the "fine" delay in the WBCAF (this block-WBCAF processor can be efficient while simultaneously performing the delay then scale form). In addition, this WBCAF is "significantly" different than the wavelet transform, not "trivially" different as the "delay-then-scale" form is (detailed in Chapter 6). This second wideband processor (the WBCAF) is desirable for imaging applications because the wavelet transforms will represent the time domain parameters while the WBCAF will represent the spatial parameters.

Some properties of the WBCAF operator have been derived [Alt, Dau, Sib, Swi, Vak]. These properties include volume, symmetry, and mapping properties. One of the intuitively acceptable properties of WBCAF processing is that it provides inherently high resolution. When the signals are "wideband" processes, the time-bandwidth product (or analogous quantities in other domains) is large and since the resolution of a signal is inversely proportional to the time-bandwidth product, the resolution of such signals is high. If the process had a small (close to unity or smaller) time-bandwidth product, then narrowband processing would be valid and wideband processing would not be required; only when the time-bandwidth product might be large will wideband processing be used.

The wideband model for systems applies when the systems are non-negligibly changing over the observation interval; the impulse response (or appropriate characterization) changes with time. When the time variation is properly characterized, then the time-bandwidth product for this characterization is also large and thus leads to high resolution in the system models. In either the wideband signal or system case, a characterization adequately describing these phenomena must have high resolution. The high resolution and high gain properties of WBCAFs require a very dense evaluation set when they are estimated to avoid missing the peaks of these functions. Thus many WBCAF evaluations are required to cover all possible hypothesized scales and translations. This large number of evaluations quickly becomes overbearing for even the fastest of resources with conventional WBCAF generation algorithms [Vai]. A fast, efficient WBCAF generation algorithm is required to accomplish the high resolution and high gain localization in the environment.

The details of the narrowband cross ambiguity function are presented in Appendix 3-B but a brief review is presented for notational purposes and for those unwilling to jump forward to reference an appendix. As mentioned previously, narrowband cross ambiguity function (NBCAF) processing has been used extensively for many applications. The NBCAF is the "normal" ambiguity function [Van2, Woo] and is defined for two signals, r(t) and s(t), as:

$$\mathbf{NBCAF}(\omega_D,\tau) = \int_{T_1}^{T_2} r(t)\, s^*(t-\tau)\, e^{-j\omega_D t}\, dt \qquad (3.14)$$

where τ is the time delay and ω_D is the frequency (Doppler) shift. The NBCAF is a two-dimensional *complex valued* surface. Often the magnitude squared of the NBCAF's surface is employed as a detection statistic [Van, Zio]. Note that the time translation operates only on the one received signal and not on the exponential function.

Returning to the WBCAF, the first form of the continuous time wideband ambiguity cross function (WBCAF) between r(t) and s(t) is formulated as:

$$\mathbf{WBCAF}(a,b) = \sqrt{a}\int_{T_1}^{T_2} r(t)\, s^*(at-b)\, dt \qquad (3.15)$$

where r(t) is the received signal and s(t) is a second, possibly the same, signal. The parameter, a, is the time scale and b is the time delay or translation. The wideband ambiguity function processor has a structure as shown in Figure 3.10.

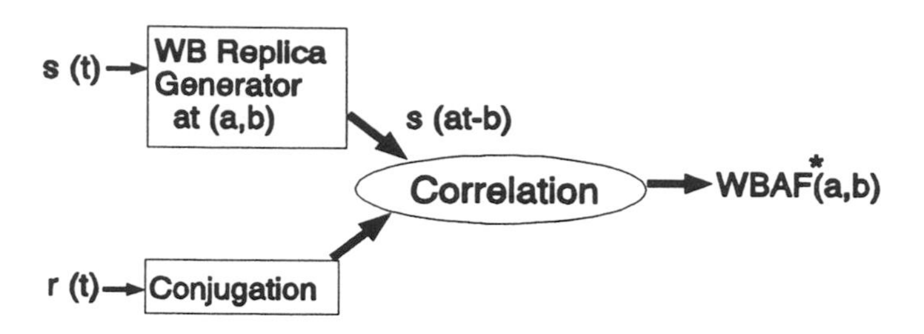

Figure 3.10: Direct WBCAF Generation

Besides the formulation of the WBCAF in equation (3.15), another WBCAF formulation is:

$$WBCAF_{not\ used}(a,b) = \sqrt{a}\int_0^T r(t)\, s^*(a(t-b))\,dt \qquad (3.16)$$

This formulation can obviously be deterministically mapped from the other formulation but will produce different ambiguity functions. There are several significant differences between these two ambiguity functions. The differences arise order of the time scaling and the time delaying (or advancing) operations. The WBCAF in equation (3.15) scales the signal first and then delays it. The second WBCAF delays first and then scales. Since the wavelet transform also delays and then scales, the wavelet transform and the WBCAF in equation (3.16) are essentially identical; thus, a second function is not required. The WBCAF in equation (3.15) is significantly different from the wavelet transform and is appropriately identified as a separate operator. Reversing the order of time scaling and delaying causes the time reference to change - should the signal be delayed in scaled time or correlation time? For passive sensor implementations on computers it is usually easier to scale a signal replica first and then delay this scaled signal; this corresponds to the first WBCAF and thus that is the chosen ambiguity function. Both formulations of the wideband ambiguity function were implemented. The discussion comparing these two ambiguity function forms is deferred to Chapter 6.

WBCAF as a Wavelet Transform

In this section WBCAFs are defined and formulated as wavelet transforms. The one-dimensional wavelet transform operator, $\boldsymbol{W}_g$, maps an $L^2(\mathbf{R})$ signal, f, as follows: $\boldsymbol{W}_g : L^2(\mathbf{R}) \mapsto L^2(\mathbf{R}\backslash\{0\}\, x\mathbf{R})$. As defined in equation (2.2) the wavelet decomposition or transform is defined as follows for a given mother wavelet, g(x):

$$(W_g f)(a,b) = \langle f, g_{a,b} \rangle = |a|^{-\frac{1}{2}} \int f(x)\, g^*\left(\frac{x-b}{a}\right) dx \qquad \textbf{(3.17)}$$

Reformulating for WBCAF processing of two unknown square integrable signals, $r_1(t)$ and $r_2(t)$, and replacing the independent variable, x, by the temporal variable, t in equation (3.15) and:

$$a = \frac{1}{s}\ ,\ b = \frac{\tau}{s}\ ,\ f(x) = r_1(t)\ \ and\ \ g(x) = r_2(t)$$

then

$$\begin{aligned} WBCAF(s,\tau) &= |s|^{\frac{1}{2}} \int r_1(t)\, r_2^*(st-\tau)\, dt \\ &= \left\langle r_1, r_{2,\frac{1}{s},\frac{\tau}{s}} \right\rangle \qquad \textbf{(3.18)} \\ &= (\boldsymbol{W}_{r_2} r_1)\left(\frac{1}{s}, \frac{\tau}{s}\right) \end{aligned}$$

This definition requires $r_2(t)$ to be an admissible function. The admissible function requirement was originally presented in reference [Sib2]. Prior to reference [Sib2] the WBCAF's usefulness was questioned because the volume under the WBCAF could not be normalized to unity and could be infinite [Alt1, Kel]. A similar argument is employed in Chapter 5 to justify a new system characterization; previous system characterizations did not properly treat energy. The new system characterization represents the energy distribution of the system.

The wavelet transform interpretation of the WBCAF has been performed by several researchers [Aus, Com, Sib1]. The following section extends the wavelet interpretation to also apply to each individual signal as well.

The $r_{2,\frac{1}{s},\frac{\tau}{s}}(t)$ functions are the time scaled and time translated versions of the mother wavelet, $r_2(t)$, where the parameter "s" denotes the time scale and the parameter "τ" represents a time translation. Any coefficient for a particular (s,τ) may be "fairly" compared or combined even across a huge range of scales (over a wideband).

Reformulation of the WBCAF with Wavelet Transforms

Previous research formulated the WBCAF as a wavelet transform but the signals themselves were still functions of time. For reasons to be subsequently stated, it may also be desirable to form a WBCAF of two signals, each of which are also represented in the wavelet transform domain (see Figure 3.11 - the components are subsequently presented). The transform domain representations can provide efficiencies for the ambiguity function processing and can also display the characteristics of the signal for other purposes as well (signal recognition, richness of the signal, etc.). The NBCAF is formulated with FFTs to enhance the computational efficiency and it also provides a spectral representation of the signal which can be used for detection, characterization and/or recognition; a similar feature extraction is provided with the WBCAF wavelet domain reformulation.

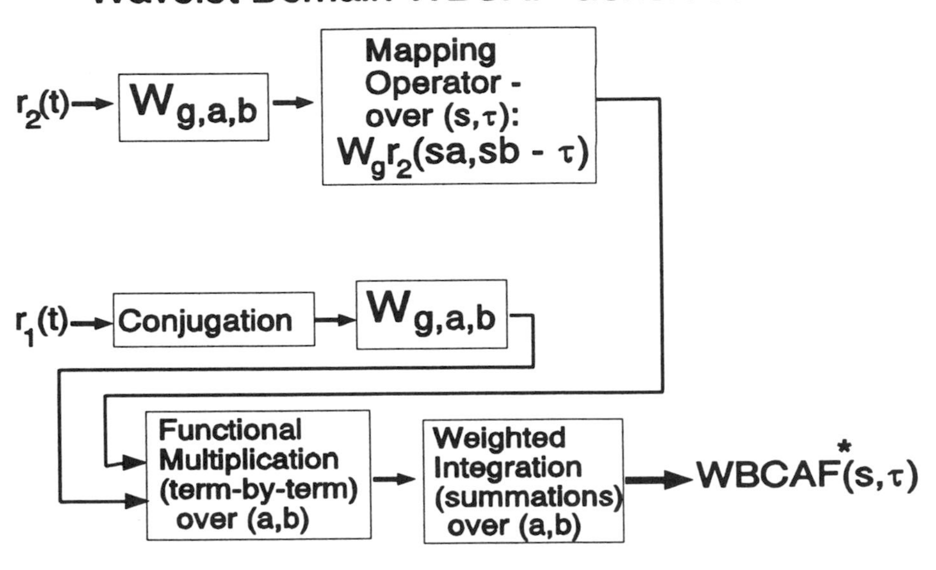

Figure 3.11: Wavelet Domain WBCAF Generation

Before deriving this new WBCAF generator, its structure is presented in Figure 3.11. Both of the time domain signals, $r_1(t)$ *and* $r_2(t)$, are wavelet transformed with respect to the same mother wavelet, $g(t)$. The scaling and translation operations of the WBCAF occur in the wavelet domain and are accomplished by translations, multiplications, and summations over two dimensions.

Since WBCAFs are wavelet transforms and vice-versa as discussed in the previous section, then why are both used? *Even though WBCAFs can be formulated as wavelet transforms and vice-versa, the WBCAFs will represent spatial parameters and the wavelet transforms will act only on time domain parameters, so both WBCAFs and wavelet transforms are used instead of just one of them.* This distinction has been used in the narrowband formulation where the Fourier transforms act on the time domain signals and the narrowband cross ambiguity functions represent the spatial parameters. This book follows the narrowband notation to exploit the analogies and to be consistent with the well accepted narrowband notation.

A theorem, the resolution of identity, was stated and proved in Chapter 2, and is applied here to the wavelet domain WBCAF reformulation. The mother mapper operator presented later in Chapter 4 significantly simplifies this derivation but it is not required (after the mother mapper operator is presented and derived, one application is to rederive this WBCAF reformulation in the wavelet transform domain).

Reformulation of the WBCAF in the Wavelet Domain - Derivation

In Chapter 2 the resolution of identity theorem was presented, derived and discussed. When g is an admissible function this theorem states the following:

$$\int_{-\infty}^{\infty} \frac{da}{a^2} \int_{-\infty}^{\infty} \langle r_1, g_{a,b} \rangle \langle g_{a,b}, f_2 \rangle \, db = c_g \langle r_1, f_2 \rangle \tag{3.19}$$

The admissibility constant, c_g, must be a well defined constant (less than infinity). From the definition of the inner product:

$$\langle g_{a,b}, f_2 \rangle = \langle f_2, g_{a,b} \rangle^* \tag{3.20}$$

so that:

$$\langle r_1, f_2 \rangle = \frac{1}{c_g} \int_{-\infty}^{\infty} \frac{da}{a^2} \int_{-\infty}^{\infty} \langle r_1, g_{a,b} \rangle \langle f_2, g_{a,b} \rangle^* \, db \tag{3.21}$$

Now the WBCAF of two signals, $r_1(t)$ and $r_2(t)$, (possibly both unknown) is defined as:

$$WBCAF(s,\tau) = \sqrt{s} \int_{-\infty}^{\infty} r_1(t) \, r_2^*(st-\tau) \, dt \tag{3.22}$$

where s and τ are the scale (compression or dilation) and translation (delay or advance) parameter.

Let $f_2(t) = \sqrt{s}\, r_2(st-\tau)$ then

$$\begin{aligned} WBCAF(s,\tau) &= \langle r_1(t), \sqrt{s}\, r_2(st-\tau) \rangle \\ &= \langle r_1, f_2 \rangle \end{aligned} \tag{3.23}$$

Representing the two signals in the wavelet transform domain with respect to the mother wavelet, g, yields two sets of wavelet transform coefficients, $W_g r_i(a,b)$, that represent $r_i(t)$ for i=1,2. So now the two received signals are both represented in the "wavelet domain."

If $f_2(t) = r_2(t)$, (corresponding to the WBCAF evaluated at $(s=1, \tau=0)$) then both sets of coefficients for the resolution of identity equation can be computed; they are just the wavelet transform of $r_1(t)$ with respect to g and the conjugate of the wavelet transform of $r_2(t)$ with respect to g. When applying the resolution of identity, the c_g constant must be taken into account, but this is a constant that may be computed in advance and since it is the same for all WBCAF points it is useless for comparing across (s,τ) hypotheses. The computations proceed as:

$$WBCAF(s=1,\tau=0) = \langle r_1(t), r_2(t) \rangle$$

$$= \int_{-\infty}^{\infty} \frac{da}{a^2} \int_{-\infty}^{\infty} \langle r_1, g_{a,b} \rangle \langle g_{a,b}, r_2 \rangle \, db \tag{3.24}$$

$$= \int_{-\infty}^{\infty} \frac{da}{a^2} \int_{-\infty}^{\infty} [W_g r_1(a,b)][W_g r_2(a,b)]^* \, db \tag{3.25}$$

which is exactly what is intuitively expected; if the two signals were the same, then their wavelet transforms should align. Note that this is the evaluation of the WBCAF at the origin of the delay-scale plane.

The general formulation of the WBCAF (instead of just its evaluation at one point) in the wavelet transform domain may be created by extending this single evaluation. The generalization comes from considering f_2 (defined previously) which is a function of s and τ instead of just a single evaluation. The key is to determine the wavelet transform coefficients of f_2 for each s and τ as a function of the wavelet coefficients of r_2.

Rewriting the second wavelet transform yields:

$$W_g f_2(a,b) = W_g r_{2,\frac{1}{s},\frac{\tau}{s}}(a,b)$$

$$= \left\langle \sqrt{s}\, r_2(st-\tau), \frac{1}{\sqrt{a}} g\left(\frac{t-b}{a}\right) \right\rangle \tag{3.26}$$

and by a change of variables:

$$= \left\langle r_2(t'), \frac{1}{\sqrt{as}} g\left(\frac{t'-(sb-\tau)}{as}\right) \right\rangle \tag{3.27}$$

then the wavelet transform coefficients of the scaled and translated version of r_2 in terms of the original wavelet coefficients of r_2 may be determined:

$$\langle f_2, g_{a,b} \rangle = \mathbf{W}_g f_2(a,b) = \mathbf{W}_g r_{2,\frac{1}{s},\frac{\tau}{s}}(a,b)$$

$$= \mathbf{W}_g r_2(as, sb-\tau) \tag{3.28}$$

Combining the previous steps by inserting equation (3.28) into equation (3.21) yields the reformulated WBCAF as a function of the wavelet transforms of the two received signals:

$$WBCAF(s,\tau) = W_{r_2}r_1\left(\frac{1}{s},\frac{\tau}{s}\right)$$

$$= \frac{1}{C_g}\int_{-\infty}^{\infty}\frac{da}{a^2}\int_{-\infty}^{\infty}[W_g r_1(a,b)]\,[W_g^* r_2(sa, sb-\tau)]\,db \qquad (3.29)$$

The WBCAF of two unknown signals may now be computed in the transform domain. Under the assumption that $r_2(t)$ is an admissible function, the WBCAF is a wavelet transform with respect to the new mother wavelet, r_2. The admissibility condition was discussed at the start of Chapter 2.

As stated above, each signal will be represented by its wavelet transform with respect to another, arbitrary, mother wavelet, g(t). Thus the WBCAF can be expressed as an integral operator acting on the two signal wavelet transforms. The significance of this reformulation is that the wavelet transforms of each signal are with respect to an arbitrary mother wavelet. Since this mother wavelet can be a priori chosen and known, then a mathematical model of its time scaling and translations will also be known. The mathematical signal model allows the scaled and translated mother wavelets to be efficiently and easily generated; thus, the wavelet transforms with respect to g are efficient and subsequently, the WBCAF generation is also efficient.

The new generator/processor reformulated the WBCAF entirely in the wavelet domain. The following sections investigate the features and structure of the new processor. Figure 3.11 displays the reformulated WBCAF generator in the wavelet domain. Note that the processor computes the conjugate of the WBCAF. This is due to the efficiency in conjugating one input instead of the transform domain representation as in equation (3.29).

This processor structure may also be compared to the NBCAF processor structure presented in Appendix 3-A, Figure 3.17. These and several other properties associated with this reformulated WBCAF are presented in the next section.

Properties of the Reformulated WBCAF

This section discusses the properties of the reformulated WBCAF and provides analogies to related processors and research. The analogies to narrowband cross ambiguity functions are fully developed. The advantages of wavelet transforming the time domain signals before any other operators act on these signals are presented.

Spatial Mapping of WBCAF

The WBCAF is defined over the *parameters of* s *and* τ; both of which are assumed to be independent of time even though the WBCAF processes time signals. For some applications the parameters s and τ can be mapped to the spatial parameters of angle (or wavenumber) and range (similar to the narrowband mapping of the delay and Doppler shift to spatial positions performed by Chestnut [Che]). Since the WBCAF may be interpreted as a spatial representation, this "spatial" wavelet transform may be formulated as an operator that acts on the two spatially separated temporal signals (which may be two received signals or one received and one transmitted signal). But instead of operating on the time domain signals, these signals are first wavelet transformed and represented in the wavelet domain (continuous or discrete). Then a new operation is constructed to act on these transformed signals to form the WBCAF. The result is that the WBCAF can now be computed entirely in the wavelet transform domain.

Comparison to the Narrowband Cross Ambiguity Function and the Fast Fourier Transform Operator

This section compares the wavelet transform domain WBCAF processor to efficient methods for generating the narrowband cross ambiguity functions (NBCAFs). An efficient method for NBCAF generation [SteS] also operates in a transform domain, the DFT domain, with the associated operator, the fast Fourier transform (FFT). An analogy is demonstrated between the WBCAF and NBCAF and their associated operators, the wavelet transform and the FFT, respectively, to demonstrate that the reformulation of the WBCAF is exactly analogous to that of the efficient generation of the NBCAF.

As presented in Section Appendix 3-A, the NBCAF of two unknown signals, $r_1(t)$ and $r_2(t)$, is defined as:

$$NBCAF(\omega_D,\tau) = \langle r_1, r_{2,\omega_D,\tau}\rangle = \int r_1(t)\, r_2^*(t-\tau)\, e^{-j\omega_D t}\, dt \qquad (3.30)$$

where τ and ω_D are the hypothesized delay and frequency shift differences between the signals. Some references refer to this function as the uncertainty function and reserve the term, "ambiguity function," for the magnitude squared of the NBCAF defined here. This author believes that the uncertainty or ambiguity exists in both the magnitude and the phase portions of the signal and that in the wideband case these cannot be separated. Since the ambiguity function should have a consistent definition for both the narrow and wideband cases, the form involving both the magnitude and phase is chosen here. The term uncertainty function is not used any further in this book. The efficient generation of the NBCAF is accomplished by formulating the generation problem in the DFT domain as is discussed in Appendix 3-A. This NBCAF processor structure is shown in Figure 3.17.

Note that this structure is identical to the processor structure derived for the WBCAF in the wavelet transform domain and displayed in Figure 3.11. The WBCAF processor and the NBCAF processor are identical except for several substitutions: the FFT operator is replaced by the wavelet transform operator, the shifting operator is extended to two dimensions and the inverse FFT operator is replaced by an operator similar to the inverse wavelet transform.

Studying the narrowband ambiguity function and its applications demonstrates application areas where wavelet transforms can be easily applied - possibly as a "drop-in" replacement for the narrowband ambiguity function.

Individual Signal's Wavelet Transforms and Nonstationarity

The reformulation of the WBCAF in the wavelet transform domain produces an advantage for signal representations and processing. The individual time series may be nonstationary. Since the wavelet transforms act on these functions immediately, their nonstationary characteristics can be maintained with a stable and efficient representation (wavelet coefficients). Thus, not only can "wideband systems" involving motion (spatial nonstationarity) be characterized by the WBCAF or spatial wavelet transform, but these spatial characteristics can themselves be characterized by nonstationary temporal signals. In the past, several researchers have attempted to characterize wideband (spatially nonstationary) systems which involved motion [Alt, Car, Chan, Kel]. These researchers assumed temporally stationary, narrowband signals (wavefields) and then decomposed them with Fourier techniques to make the problem mathematically tractable. These narrowband and stationary signal assumptions are no longer sufficient to characterize systems with non-negligible motion relative to the motion of the wavefield. Thus the individual wavelet transform maintains the nonstationary temporal signal characteristics and allows it to be utilized for characterizing spatially nonstationary systems.

This aforementioned simultaneous treatment of nonstationarity in space and time demonstrates a significant generalization of wideband space-time processing. The wavelet transforms are in both space (the WBCAF) and time (the individual signal's temporal wavelet transforms). However, all three wavelet transforms, including the WBCAF wavelet transform, operate on the time parameter. As Einstein points out, space and time are interchangeable, it just depends upon your point of reference. In these cases, the motion between the signals and the sensors creates several reference frames.

For the reformulated WBCAF, the wavelet domain shifting in both dimensions in equation (3.29) represents the scale and translation of the one sensor's wavelet transform. This scaling and translation act on the *spatial only* variables, (s, τ). Each evaluation of the WBCAF at a (s, τ) value represents the degree of similarity between a scaled and translated version of one received signal and an unmodified version of the second received signal. However, since it was assumed that the wavefield was created/modified by sources/structures that are in the view of both sensors, then the degree of similarity between the signals represents the alignment of these signals. Under additional assumptions these "degrees of similarity" can be mapped to spatial positions to create a *spatial image*. As stated earlier, the original wavelet transforms are in the time domain and so is the WBCAF wavelet transform; thus the time and space dimensions are nonseparable in this processor and a multidimensional space-time wavelet transform has been constructed.

This space-time wavelet transform can characterize both the temporal nonstationarities of the signal (the two functions in the (a, b) plane) and the spatial structure (the WBCAF in the (s, τ) plane) that created/modified the received multidimensional wavefield. The key to the new multidimensional wavelet transform in equation (3.29) is that it maps two two-dimensional functions to one two-dimensional function so that the dimensionality does not increase. Original attempts to apply wavelet transforms to the space-time processing problem had wavelet transforms operating on time signals and attempting to return to time signals. These attempts failed due to the dimensionality continually increasing as more wavelet domain operators were added.

Possible Efficiency of Reformulation

One advantage of the WBCAF reformulation with wavelet transforms is its efficient implementation structure. First, when both of the signals in the WBCAF are unknown, then this multidimensional wavelet transform processor will allow an arbitrary *known* mother wavelet to decompose the signal into the coefficient domain. Then the scaling operations in this domain are mapped to shifts. When fine scales are required this shifting operation may be more efficient than the multirate filtering operations [Cro] - details of an efficiency analysis are not performed in this book and are highly sensitive to the chosen architecture. For high resolution processing (often associated with wideband processing) the scale shifts may be very fine. For fine sample rate changes the perfect reconstruction method becomes unbearably inefficient. Thus the new formulation can effectively generate the scaled replicas easily in the transform domain.

In many situations more than two sensors are involved and must be cross correlated. The multiple WBCAFs may be generated in the wavelet domain but each signal must only be transformed once. In addition each shifted wavelet transform may be used multiple times and only generated once. The only operations required for additional sensors are the multiplications in the wavelet domain and the subsequent integrations or sums.

Besides these "direct" advantages, if an efficient wavelet domain representation can be constructed (forming these efficient representations is discussed later in this chapter), then the wavelet domain WBCAF generation will become even more efficient.

Before a comparison of the wavelet domain WBCAF generator to traditional wideband processors can be made, the traditional WBCAF generators must be presented. The "traditional" processors are the direct WBCAF generators described in Appendix 3-A and the multirate filters discussed in [Cro, Vai]. WBCAFs have been computed for a long time. When one of the signals is known a priori then the scale and translated replicas of the known signal may be precomputed "off line" and the WBCAF may be computed by parallel correlators using these precomputed replicas.

Unfortunately, in many applications, both of the signals are a priori unknown and the replicas of one signal must be computed in real time. Wideband replica generation is difficult in the digital domain. If the signals are stored in an analog format and can be sampled at continuously varying sample rates, then the replicas can be easily formed. Usually both of the received signals are digitized near the front end processing; this has many benefits [Bos2]. For WBCAF generation, digitized wideband replicas of digital signals requires a resampling or similar sampling rate conversion process.

If high resolution WBCAFs are required, then the resampling or sample rate conversions are in very fine steps. For signals with time-bandwidth products in the thousands the sampling rate may change by only one part in thousands. When this is required, the upsampling and interpolation rates also increase by thousands. For perfectly reconstructed replicas at each of these time scales, thousands of different filters must also be constructed. Then thousands of correlations must occur with these replicas. Obviously, for any reasonable size signals and any fine resolutions, the computations quickly exceed even the fastest of machines. The bottleneck of direct generation is these fine scaled replicas (see Figure 3.12). Inexact replica generation can be used as well as a series of coherent narrowband processors but both suffer from errors which are not easily bounded and still require excessive computations.

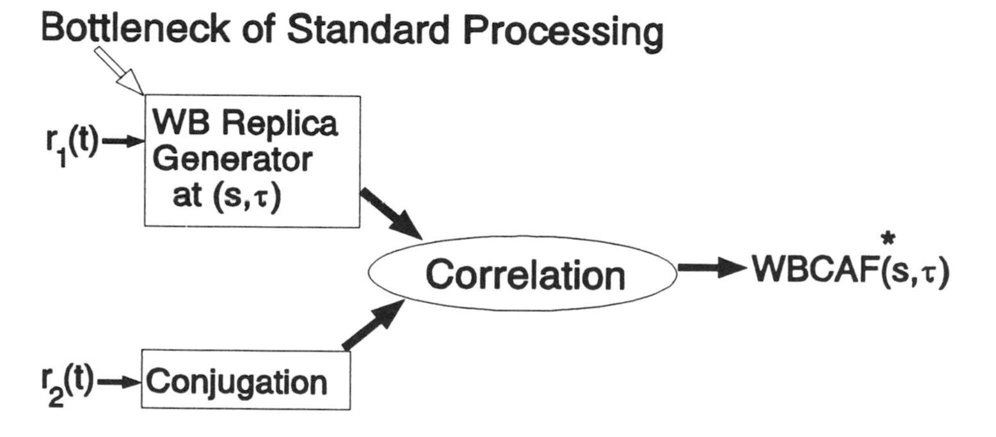

Figure 3.12: Bottleneck of Direct WBCAF Generation

The proposed algorithm operates in the wavelet transform domain. The scaled and translated mother wavelets act as precomputed "replicas" for both signals. Once the wavelet transforms have been computed, the algorithm fixes the one set of coefficients (possibly conjugates them first) and then operates on the other set. The operations consist of shifts in the scale parameter; the same shift occurs for all translation values. This scale shift is essentially the generation of a scaled replica in the transform domain, but it is much more efficient than those available in the time domain. The scale shifts are followed by time translations in this shifted coefficient space. Now the scaled and translated replica in the coefficient domain has been generated. Since all of these shifts will be known in advance, the coefficients may be mapped in parallel to their respective new locations in the transform domain.

The replica generation process has been simplified to a series of shifts in the wavelet transform domain. The wide ranging resolutions in both parameters are built into the transform domain. Oversampling in the transform domain also occurs for almost every admissible mother wavelet; this may help provide finer resolutions.

The correlation process is a term-by-term multiplication of the two sets of wavelet coefficients followed by a weighted integration. Note that the weighting is only over the scale domain though, so that the integration in the translation domain is not weighted.

The new WBCAF generation may be interpreted as a cross wavelet spectrum and a simple, efficient inversion formula. This is similar to the narrowband cross ambiguity function generation in terms of the cross spectrum. The efficiency of this approach is due to the approximation of the cross spectrum by the term-by-term product of the individual signal FFTs [SteS].

The problem with stating specific efficiency values is that the efficiency is not just a function of the "new" processor versus the "old" processor. Efficiencies are also a function of the underlying algorithms (such as the efficient wavelet domain representation with multiple mother wavelets) that are integrated to construct the "old" and "new" processors, the implementation architecture and communication techniques, the resolution requirements and the chosen "old" algorithm. This extensive problem of quantifying the efficiencies is deferred to future research.

Cross Wavelet Transforms and Signal Commonalities

The wavelet domain formulation of the WBCAF produces a structure that is exploited several times later in this book. Two wavelet transforms are processed together (crossed) to form a new ambiguity function or wavelet transform. Thus, by cross processing two wavelet transforms a new *cross* wavelet transform is formed (WBCAF). Later, in Chapter 4, this operation is generalized and referred to as the mother mapper operator.

When the two wavelet transforms being processed are with respect to the same mother wavelet, then the *cross wavelet transform indicates the extent of commonality between the two original wavelet transforms or signals.* The magnitude of the cross wavelet transform can be used to detect the presence of common signals or localize the source of these common signals. The commonality can exist for several different reasons and each application can lead to different justifications for this commonality. One example of the commonality between two wavelet transforms or signals is the case in which two receivers observe a common signal source (emitter, reflector, reradiator, etc.). The magnitude of the cross wavelet transform will have a peak that indicates this commonality. Multiple common signals can be observed and cross processed as well; these multiple common signals in each received signal will produce multiple peaks in the cross wavelet transform. Several applications exist that exploit this form of cross processing.

For processing the signals received at two spatially distinct receivers, the cross wavelet transform can extract the portions of these signals that are common to both received signals (signal portions that are "viewed" by both sensors). First the wavelet transforms are formed - the time domain signals are wavelet transformed with respect to an arbitrary mother wavelet. Consider a particular case. The signal received at sensor 1 is shown in Figure 3.13 and the received signal at sensor 2 is also a sum of two FM signals but at different delays and scales (corresponding to the different spatial sources).

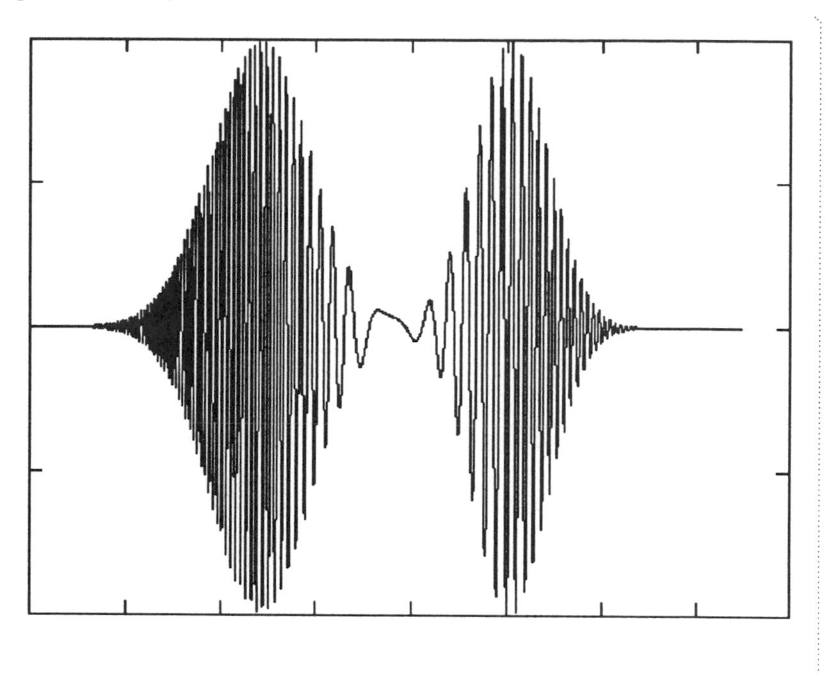

Figure 3.13: Sensor 1 Signal - Two FM's

These two time domain sensor signals can be cross wavelet transformed to create a spatial characterization. See Figure 3.14. These two wavelet transforms are cross processed (with the wavelet domain WBCAF generator) to create a new wavelet transform (or WBCAF). See right of Figure 3.14. The distinctive feature of this wavelet transform is that it is in the space domain; the first two transforms were in a scale and translation domain related to the temporal parameter. Now the scale and translation of this cross wavelet transform represent spatial parameters such as angle and range or range and velocity.

Cross Wavelet Transform-Maps Time to Space

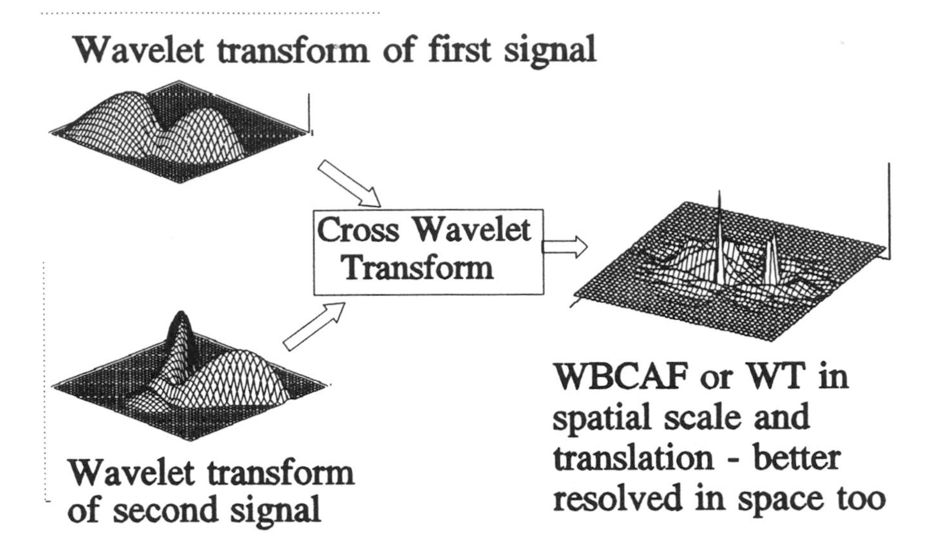

Figure 3.14: Cross Wavelet Transform Example

Both the first and second signals being transformed in Figure 3.14 are sums of two quadratic and linear FM signals (a different FM signal each of the two moving sources) as were shown in Figure 3.13. The wavelet transforms of these signals shown in Figure 3.14 are with respect to the Morlet mother wavelet discussed in Chapter 1 (the cross processing makes this intermediate mother wavelet arbitrary). The cross wavelet transform maps the two received signals into a spatial wavelet transform or WBCAF. Note the improvement in resolution and gain in this spatial domain wavelet transform. This cross wavelet transform provides a time-scale (time-frequency) representation as an intermediate feature which could be very useful for some data fusion or neural net processing algorithms; it also provides the desired final result, a spatial characterization.

As a different example of this commonality, consider two output signals from two different systems that had the same input signal. The output signals can be wavelet transformed and then a cross wavelet transform can be created from these two individual transforms. This cross transform then represents the commonality (and the differences, of course) between the two systems. These measures of similarity can be employed in feature extraction or modelling schemes.

Later in this book, a new operator, the Mother Mapper Operator will be derived. This new operator has the same form as the wavelet domain WBCAF generator. But in this operator the commonality is between wavelet transforms with respect to different mother wavelets; details are subsequently provided.

Appendix 3-A: Wideband/Narrowband Ambiguity Functions: Assumptions, Tradeoffs and Efficiencies

Ambiguity functions have many forms including combinations of wideband/narrowband, auto/cross and can be formulated with Fourier, FFT and wavelet transforms. However, each form and each signal model requires a variety of assumptions; when the appropriate assumptions are invalid, the corresponding processor is invalid. The wideband ambiguity function is sometimes incorrectly formulated and implemented due to invalid assumptions and models. This appendix identifies the assumptions for each ambiguity function processor and its associated advantages, efficiencies and tradeoffs when compared to other ambiguity function processors. Finally, ambiguity function implementation techniques with wavelet, Fourier and FFT transforms are discussed, highlighting possible invalid assumptions.

Recently, wideband ambiguity function processing has been more extensively researched and discussed due to its relationship to the wavelet transform. An earlier section in this chapter formulated the "Full-WB" ambiguity function process as a wavelet transform. The problem with processing these fully modulated signals (signal at the center frequency) is the enormous number of samples that are required to represent the signal. Since all subsequent processing must operate on these large sample sets, the processing requirements are likely to become too intensive to be practical. In addition, the operation of time scaling (resampling) is very inefficient for any received signals due to the sampling rate changes that are required.

Many methods exist for computing WBCAFs; however, all of them which operate on digital data are extremely computationally intensive. The estimation of one point of the WBCAF requires the generation of a scaled and translated wideband replica of one signal and the correlation of this replica with the other signal - just as it is defined. FFT based algorithms utilized in narrowband ambiguity function evaluation [SteS] cannot be used due to the narrowband conditions stated in Chapter 3 being invalid for wideband signals. The theoretically correct approach is to employ multirate perfect reconstruction filters to generate the scaled and translated replicas and then use fast correlation methods to evaluate the WBCAF for that particular scale and translation. Unfortunately, although multirate perfect reconstruction filters are efficient for select cases of sample rate changes, they become extremely inefficient for high resolution sample rate (scale) changes. The upsample rates must be very high to achieve the high resolution and then the decimation filters do not overlap much because they must be formed at every down sample rate. The density of prime numbers makes this approach unacceptable because an entirely new filter must be designed for each prime number. This will produce a large bank of unique filters with very little commonality (and thus decimation) between the filters. Since the efficiency of the multirate perfect reconstruction filters is proportional to the degree of commonality between the filters, it is obvious that this approach will not be much more efficient than direct generation (not multirate) of the replicas. Thus, even with the latest approaches to filtering, the generation of high resolution WBCAFs is not much better than direct evaluation. Note that signals requiring a wideband representation to achieve gain are most likely to have many samples that need to be processed coherently; this adds complexity to the already impractical WBCAF generation.

Ambiguity Function Forms and Associated Assumptions

Several ambiguity functions exist and there are assumptions related to each ambiguity function processor that depend upon the signal characteristics. In general, the more assumptions that are valid, the more efficient the ambiguity function processing becomes. However, if an ambiguity function processor is used without satisfying its related assumptions then the ambiguity function that is formed is invalid and may not be representative of the signal characteristics. Thus, for valid processing the appropriate ambiguity function must be chosen to correspond to the minimum set of assumptions that will always be satisfied; otherwise, incorrect results may be generated.

The four ambiguity functions that are to be considered are the Full Wideband, Analytic Signal Wideband, Full Narrowband, and the FFT Narrowband ambiguity functions. Other ambiguity functions also exist that concentrate on efficient estimation of the direct ambiguity functions [Gli, Ric2]. The Full Wideband (Full-WB) ambiguity function is the "normal" wideband ambiguity function [Alt, Kel, Sib]. The **Full-WB ambiguity function** is the most general ambiguity function since all of the other ambiguity functions are approximations of this ambiguity function. It is formulated as:

$$Full\text{-}WB(a,b) = \sqrt{a}\int_0^T r(t)\, s^*(at-b)\, dt \qquad \textbf{(3.31)}$$

The Full-WB ambiguity function correlation process is identical to the wavelet transform [Chai, Com, Sib]; however, in the wavelet transform one signal is replaced by an a priori known "mother wavelet" instead of being a general signal. The Full-WB ambiguity function processor has a structure as shown in Figure 3.12.

The **Analytic Signal Wideband (AS-WB) ambiguity function** employs the analytic signal model of the wideband signal as follows:

$$AS\text{-}WB(a,b) = \sqrt{a}\int_0^T r_{LP}(t)\, s^*_{LP}(at-b)\, e^{j2\pi f_c(1-a)t}\, dt \qquad \textbf{(3.32)}$$

where $s_{LP}(t)$ is the baseband equivalent portion of the original signal (the LP subscript denotes "low pass") at the center frequency, f_c, and $r_{LP}(t)$ is the baseband equivalent of the received signal which was also assumed to be centered at f_c. The baseband equivalent portions are the amplitude and phase envelopes presented in equation (3.1). If the analytic model is valid for the wideband signals, then this processor can be employed to operate on *both* the envelope and the center frequency exponential.

The AS-WB ambiguity function processor has a structure as shown in Figure 3.15 (essentially it is the Full-WB processor at baseband with a different frequency shift for each scale value). The phase uncertainty that is typical of many local oscillator mixers may be unacceptable in this processor.

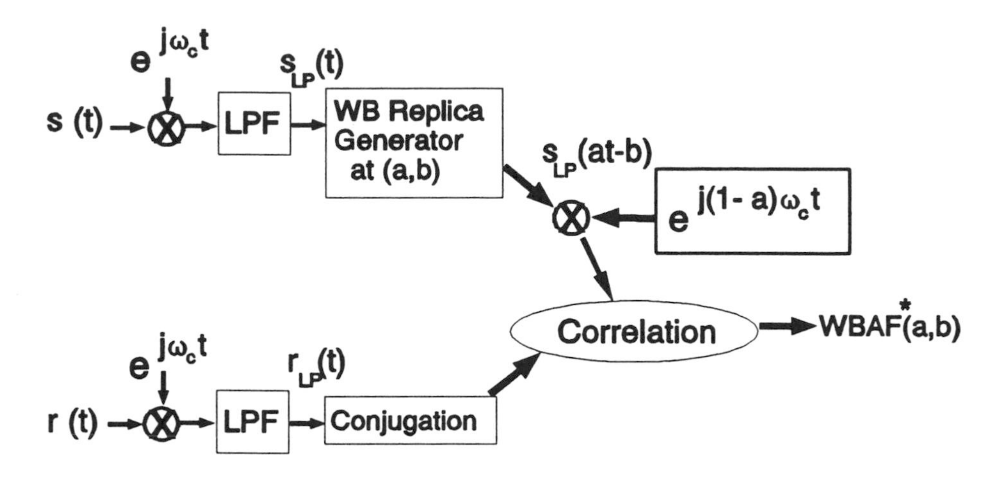

Figure 3.15: Analytic WBCAF Generation

When a received signal is time scaled to form a scaled replica signal (these are required for wideband ambiguity functions), the time scaling must act on the carrier as well as the envelopes. Thus, a received signal cannot be basebanded and then scaled without accounting for the scaling of the center frequency. Equation (3.1) can be time scaled to create the signal:

$$f(st) = a(st)\, e^{j\phi(st)}\, e^{j\omega_c st} = a(st)\, e^{j\phi(st)}\, e^{j(1-s)\omega_c t}\, e^{j\omega_c t}$$
$$= a(st)\, e^{j\phi(st)}\, e^{j\omega_D t}\, e^{j\omega_c t} \qquad (3.33)$$

where ω_D is the Doppler frequency shift. Thus, to appropriately time scale this analytic signal, the envelopes must be time scaled and a different Doppler shift must be applied for each different scale ($\omega_D = (1-s)\,\omega_c$). For these analytic signals *a combination of both time scaling and frequency shifting must be applied.* The amplitude and phase envelopes must be time scale and the carrier frequency must be Doppler shifted.

Under the narrowband conditions, the phase variation caused by the modulating envelopes is assumed to be negligible when compared to the variation of the center frequency exponential. Thus, for small scale values the envelope's phase variations are still negligible when compared to the center frequency, but the scaled center frequency produces the Doppler frequency shift that is non-negligible. So, *for the narrowband case and small scale values the time scaling is approximated by a frequency shift*. This condition is exploited in Appendix 3-B for computing the narrowband replicas and the narrowband ambiguity functions.

The **Full Narrowband (Full-NB) ambiguity function** is the "normal" narrowband ambiguity function [Van, Woo]:

$$Full\text{-}NB(\omega_D,\tau) = \int_0^T r(t)\, s^*(t-\tau)\, e^{j\omega_D t}\, dt \qquad \textbf{(3.34)}$$

The Full-NB ambiguity function processor has a structure as shown in Figure 3.16.

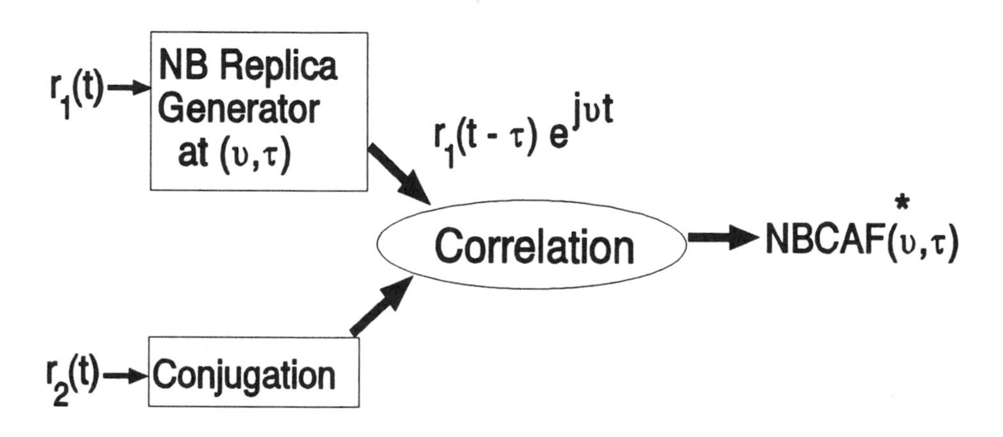

Figure 3.16: Full NBCAF Generation

The **FFT Narrowband (FFT-NB) ambiguity function** employs the FFT algorithm to enhance the efficiency of the implementation [SteS]:

$$\begin{aligned} FFT\text{-}NB(n,m) &= FFT\text{-}NB\left(nT_s, m\left(\frac{F_s}{N}\right)\right) \\ &= \boldsymbol{FFT}_n^{-1}\{[\boldsymbol{FFT}_k^*(r(kT_s))][\boldsymbol{CS}_m \boldsymbol{FFT}_k(s(kT_s))]\} \end{aligned} \qquad \textbf{(3.35)}$$

where $\boldsymbol{FFT}_k$ is the fast Fourier transform over the index k, $\boldsymbol{FFT}_n^{-1}$ is the inverse fast Fourier transform back to the index n, and $\boldsymbol{CS}_m$ is a circular shift operator which shifts the transform by m bins. The FFT-NB ambiguity function processor has a structure as shown in Figure 3.17.

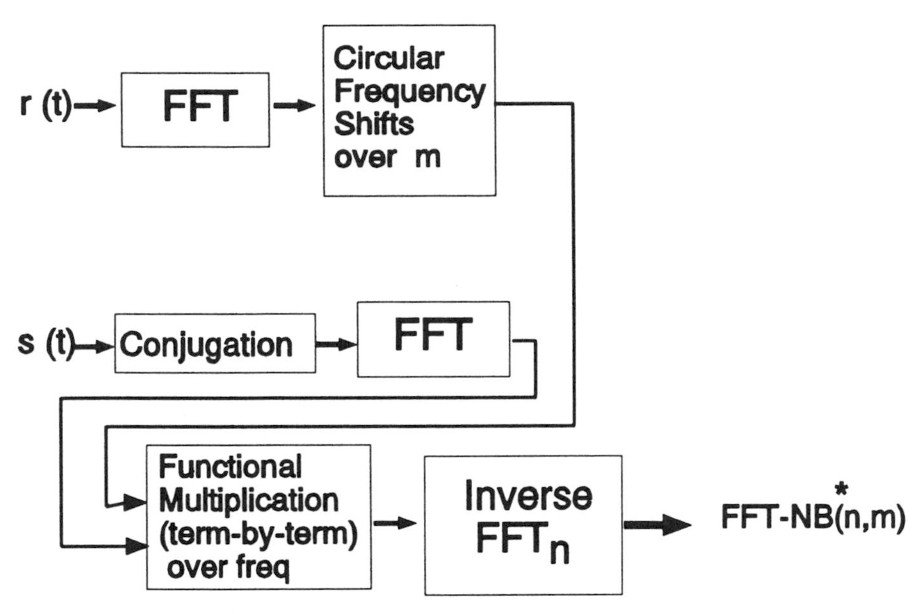

Figure 3.17: FFT Narrowband Cross Ambiguity Function Generation

Envelope Analytic Signal WBCAF Form

Another form of the WBCAF exists. This form, the envelope AS-WBCAF, is the AS-WBCAF with the exception of the frequency shifting term. The envelopes are extracted from the carrier exponential to form a lowpass equivalent signal (analytic signal). Then the lowpass equivalent of the other signal is formed. The analytic signal envelopes are then processed by the Full-WB ambiguity function (wavelet transformed) to supposedly form the equivalent of the Full-WB; this processing is shown in Figure 3.18. This processor is *invalid*. The invalidity of this processor is subsequently demonstrated.

Processing only the envelopes (signals mixed the their baseband equivalents) reduces the number of samples and the sampling rate. First, the analytic signal model is invalid for some wideband signals as discussed earlier in this chapter.

Under the assumption that the analytic signal representation is valid, another reason for the invalidity of this new processor is that it improperly scales the time argument in the replica signal. Remember that when the signal is scaled by the physical reflection process or other phenomenon, it is at the modulated frequency, thus the center frequency exponential is scaled. As in the AS-WB ambiguity function processor, each different scale must have an associated different Doppler frequency shift. The new processor does not scale the center or carrier frequency exponential; thus, it is invalid. When the Doppler shift is not included the effect on the wideband ambiguity function is significant. Peaks of the true WBCAF can become near zeros while zeros of the true WBCAF can become peaks.

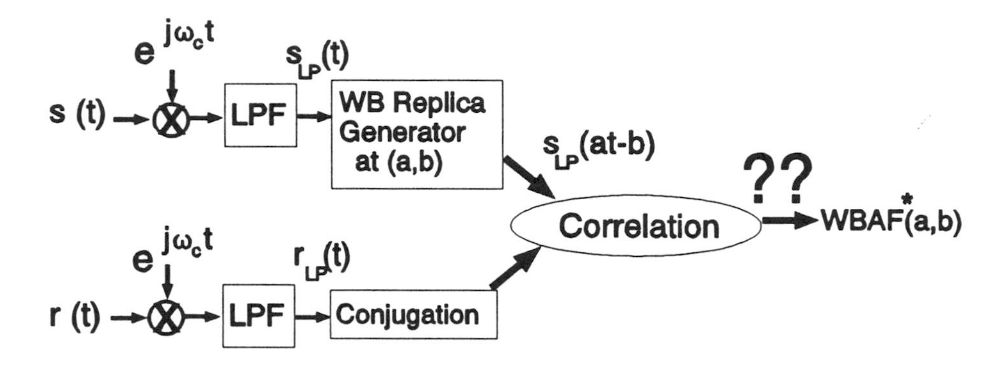

Figure 3.18: Invalid Lowpass WBCAF Processor

Besides the possible continuous time wideband processor problems, additional problems arise when discrete time implementations are attempted. The time scaling operation for discrete signals requires the resampling or extreme oversampling of signals. Multirate digital signal processing [Vai] addresses the time scaling problem; however, when the ambiguity function resolution requirements dictate fine resolutions the processing requirements quickly exceed any available processing assets. The multistage filtering efficiencies only exist for a small number of the required resampling rates and this efficiency is not available for many resampling rates that are required. Thus, huge filters must be implemented at high processing costs. Even when the processing burden is accepted, other conditions must be satisfied to form correct wideband ambiguity functions with discrete signals. The fast Fourier transform (FFT) representations are sensitive representations for wideband signals and if used outside their valid range may lead to invalid wideband ambiguity functions being formed.

The aliasing problems with discrete time and frequency domains lead to replica signals which are invalid. The problem is that the signal representations are repeated in both the time and frequency domains. When a delayed signal replica is formed the resulting time series is an aliased version. The time signal can have support in areas where the original signal did not. For a discrete frequency representation, a delay can be implemented by proper phase shifts, but these phase shifts are cyclic and can exceed 360 degrees. The delayed signals then become "aliased" versions; the end of the signal wraps to the beginning of the signal. This aliased replica is invalid. The misaligned replica support will obviously lead to invalid ambiguity function results. Zero padding can be incorporated to zero the end of the signal; however, when scale factors become significant, the zero padding can become the majority of the signal - an undesirable result. Generally, discrete frequency representations are not compatible with generating wavelet transforms.

Full-WB ambiguity function processors (continuous time wavelet transforms) have been proposed which employ FFTs as the initial signal representations. Besides the aforementioned aliasing problems when delaying signals, another problem arises with discrete FFT representations of wideband signals. The resolution requirements of the wideband ambiguity function dictate the required signal representation resolution. When a wideband or nonstationary signal is represented with an FFT its frequency and time resolutions are the same across the entire band and the time localization information is contained entirely within the phase. The same resolution at the lowest frequency and the highest frequency may lead to sensitive requirements.

Consider a signal that has components in the 10 Hz region that require 1 Hz resolution and also 10 kHz components that require a 1 kHz resolution, then the FFT produces a representation which must have 1 Hz resolution at all frequencies (the lower frequencies are extremely oversampled). Constant Q or constant fractional bandwidth implementations may more naturally (and more efficiently) represent wideband signals and will also lead to less sensitive requirements; however, these representations also have increased dimensionality (2D instead of 1D representations). Multi-band or multi-octave FFT processing is a valid alternative but coherency must be maintained over all of these bands to achieve the desired gain and resolution. Thus, to achieve high resolution in the wideband ambiguity function (or wavelet transform) with the FFT, the delays and scales must be very small and the signal representation (at all frequencies) must be at least the desired resolution of the wideband ambiguity function. These stringent requirements would make FFT implementations of wideband ambiguity functions inefficient and undesirable.

In conclusion, for multiple octave or wideband signals, the analytic signal is an invalid representation. Quadrature sampling schemes [Bos1, Vai] will not accurately represent these signals; the decimation to reduce the number of samples cannot be achieved. The "real" representation is the only valid representation. The sample rate should exceed the highest frequency component in the signal at the carrier frequency. The processing will act on these real samples only and not some "analytic" complex sample set. The Full-WB ambiguity function implementation should be employed with real input signals for most wideband signals.

Appendix 3-B: Narrowband Ambiguity Function Theory

The theory of narrowband cross ambiguity functions (NBCAFs) originates with statistical correlation processing of stochastic signals. The uncertainty associated with locating the peak of a cross correlator was interpreted as an "ambiguity" in this representation and thus its name. NBCAFs were also referred to as uncertainty functions for the same reasons.

NBCAFs represent the similarity between one signal and many delayed and frequency shifted versions (*replicas*) of another signal. The "similarity" is measured by a single correlation value for each frequency shift and translation. When the two signals are the same, an auto-ambiguity function is formed; when the signals are different, a cross ambiguity function is formed. Under another specific set of assumptions (regarding both the signals and the noise), certain forms of the NBCAF are optimal position and velocity (state) estimators [Alt, Van]. In addition, when the environment can be characterized by a "scattering function," the NBCAF can be employed to design a signal best suited for that particular environment [Alt, Van, Zio].

Narrowband Cross Ambiguity Functions

The narrowband cross ambiguity function (NBCAF) of two unknown signals, $r_1(t)$ and $r_2(t)$, is defined as:

$$NBCAF(\omega_D,\tau) = \langle r_1, O_H(\omega_D,\tau)\, r_2\rangle = \langle r_1, r_{2,\omega_D,\tau}\rangle$$

$$= \int r_1(t)\, r_2^*(t-\tau)\, e^{-j\omega_D t}\, dt \qquad (3.36)$$

where τ and ω_D are the hypothesized delay and frequency shift differences between the signals. The expression $r_{i,\omega_D,\tau}$ is defined by an operator on the r_i signal (for the mathematically inclined, this operator is related to unitary operators and representations of the Heisenberg group (justifying the H subscript) [Aus, Chai, Sch, Sib]):

$$r_{i,\omega_D,\tau} = O_H(r_i(t)) = r_i(t-\tau)\, e^{-j\omega_D t} \qquad (3.37)$$

Using group theoretic techniques the NBCAF may be derived as the limiting case of the WBCAF [Chai]. This is intuitively expected because in physical systems involving motion of either the sensors or the select portions of the environment, the NBCAF only approximates true time scaling with frequency shifts. Thus, any new results that are applicable for the WBCAF will be applicable to the NBCAF but not necessarily vice-versa.

Under a specific set of assumptions, the NBCAF naturally arises in detection and estimation theory as the maximum likelihood sufficient statistic for an environment consisting of a fluctuating point reflector in stationary Gaussian noise [Aus, Car, Sch, Ows, Van, Zio]. Thus, the NBCAF may be used to optimally estimate the environment (or a select portion of it). However, the set of assumptions assumed for this statistical estimator must be satisfied.

The aforementioned optimal estimator's assumptions include wide-sense stationarity over an observation interval that is significantly longer than the non-negligible width of the main lobe of the time correlation of the two signals being processed. If the inverse of the bandwidth, (1/B), is approximately the width of the time correlation's mainlobe and T is the observation interval, then the narrowband estimator requires:

$$T \gg \left(\frac{1}{B}\right) \tag{3.38}$$

The WSS condition and the long observation interval are required so that differential time and frequencies between the two signals being processed can be estimated and so the statistical formulation is valid. Essentially, these assumptions allow the support of the two-dimensional correlation function (with two time arguments as the axes) to be concentrated on a diagonal line (one-dimensional). This one dimensional line then represents the differential delay between the two signals. This condition can only occur if the signal's time-bandwidth product satisfies the narrowband condition in equation (3.10). The spatial equivalent of this condition was presented as Figure 3.4.

The efficient generation of the NBCAF is accomplished by formulating the generation problem in the DFT domain [SteS]. The DFTs of the input signals are computed with FFTs. The efficiency is accomplished by using partial sums several times rather than recomputing those partial sums (similar to FFT ideas) [Bos1]. The resulting processor is:

$$NBCAF^*(\omega_D, \tau) = \mathbf{FFT}^{-1}\{[\mathbf{FFT}^*(r_1)][O_{\omega_D}\mathbf{FFT}(r_2)]\} \tag{3.39}$$

where the operators, $\mathbf{FFT}$ and $\mathbf{FFT}^{-1}$, are any of the standard forward and inverse FFT operators, respectively [Bos]. The shifting operator, O_{ω_D}, acts in the discrete Fourier transform domain by circularly shifting the DFT of r_2 by ω_D. The inverse Fourier transform then reintroduces the τ variable as its time variable. The result is a two dimensional function in (ω_D, τ), the NBCAF. This NBCAF generator is displayed in Figure 3.19.

A very important application of the NBCAF is to describe the output of a narrowband correlation receiver. These details are provided in many references [Alt, Van, Zio]. NBCAFs are also integrated with narrowband scattering functions to describe stationary systems but the process of scattering is not discussed until related wavelet theory applications are detailed in Chapter 6.

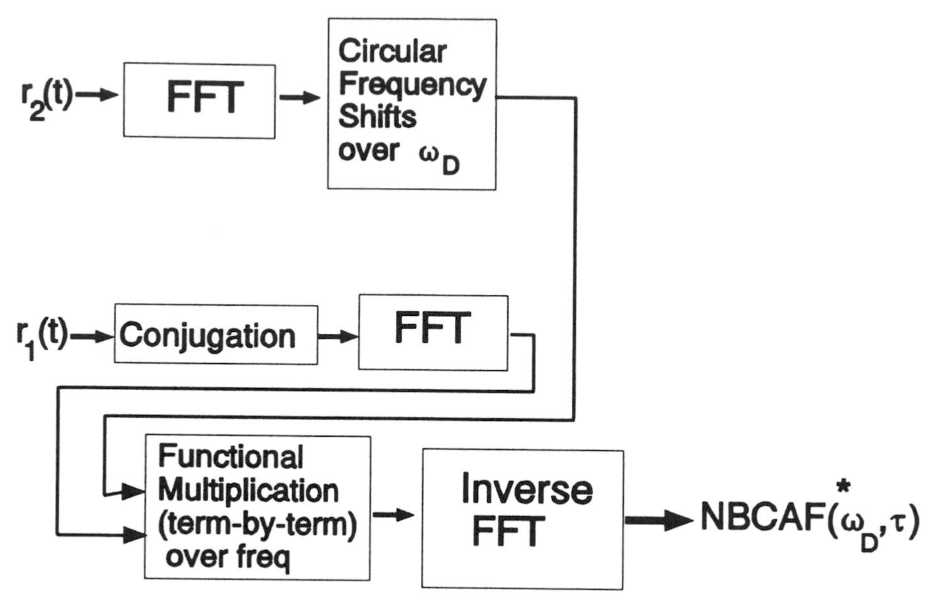

Figure 3.19: FFT Narrowband Cross Ambiguity Function Generation

Chapter 4: Wavelet Theory Extensions and Ambiguity Functions

Introduction

This book began by presenting general, continuous-time wavelet theory. Then general discrete wavelet transforms (both CTWS and DTWS) were presented. Special subsets of the discrete wavelet transforms were then presented, which included the multiresolution and the orthogonal wavelet transforms. Since these special wavelet transforms are extensively employed, this chapter compares these transforms to the less constrained wavelet transforms. A new operator is derived that improves the efficiency and interpretation of the unconstrained wavelet transforms and, in general, extends wavelet theory. Improvements and extensions are achieved for wideband correlation receivers and the wideband ambiguity functions.

This remainder of this book emphasizes the general, nonorthogonal, wavelet transforms and the operations associated with them. Since physical insights already exist for correlation processing (and ambiguity or uncertainty functions), correlation and ambiguity functions are related to wavelet transforms and efficiently implemented with them. Several different wavelet transforms are compared to help determine the type of wavelet transform that is appropriate for a specific application.

The new operator that is introduced, the mother mapper operator, maps a wavelet transform with respect to one mother wavelet to a new wavelet transform with respect to a different mother wavelet. The interpretation of this operator and its actions are visualized and the efficiency improvements and potential applications of this operator are identified.

Multiresolution/Orthogonal Wavelets versus Unconstrained Wavelets

Multiresolution wavelet transforms were briefly discussed in Chapter 2. These wavelet transforms can be extremely efficient for operations on sampled and multi-octave signals and have been successfully implemented in several applications. Multiresolution wavelet transforms yield specific (and, with additional constraints, unique) mother wavelets due to the constraints imposed upon them. Figure 4.1 identifies some of these constraints and the paths taken to create two specific multiresolution orthogonal wavelet decompositions. These multiresolution orthogonal transforms are appropriate for image analysis and some signal analysis but may be augmented by the general wavelets to create further efficiencies and in applications where mother wavelet constraints are undesirable. By using general, unconstrained, non-multiresolution (unconstrained) wavelet transforms, the mother wavelets are required to only be admissible functions (see Chapter 2). The freedom in choosing a mother wavelet can then be exploited to better match the signal, image, or system characteristics. A better match can lead to more efficient representations of the signal's energy or the system's operations.

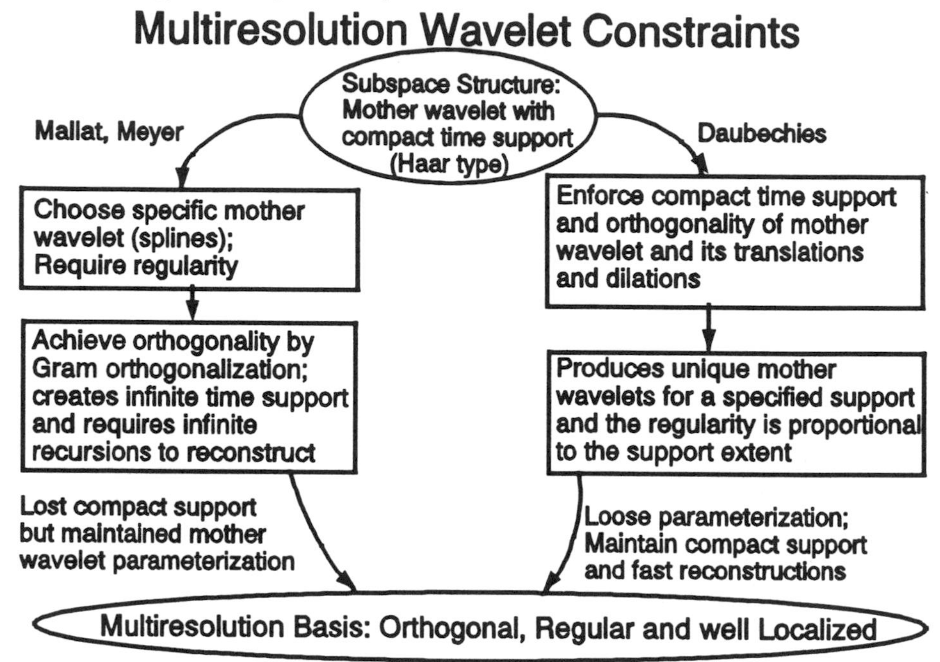

Figure 4.1: Multiresolution Orthogonal Wavelet Constraints

Consider the multiresolution orthogonal wavelet transforms. A detailed comparison between the multiresolution orthogonal wavelet transforms and the "unconstrained" wavelet transforms could not be performed until this point in the book because the comparison would require an understanding of most of the previously presented concepts. Primarily, the comparison will concentrate on the resolution and gain properties of the transforms. The resolution will be measured with the wideband ambiguity function.

The multiresolution (orthogonal, biorthogonal, or PR-QMF) wavelet transforms all exploit a dyadic lattice. The scale parameter and the translation parameters must change by integer powers of 2. As will be discussed later, scale factors that are powers of 2 are not amenable in many applications. However, for discrete implementations, scales and translations which are powers of two are desirable for efficiency reasons. Non-dyadic wavelet transforms do not have this property and typically require finer scaling that leads to inefficiencies. Non-dyadic wavelet transforms can easily operate on discrete data but they are not as efficient as the multiresolution wavelet transforms.

The primary concept behind multiresolution wavelet transforms is to divide and conquer *without any redundancy*; a continual pyramidal division process chops the signal into pieces (filter outputs). See Figure 2.15. Each piece, or filter output, represents a portion of the original signal's input spectrum; the spectrum is divided into constant-Q bands over adjacent octaves (frequency bands are split into two parts by each filter) due to the dyadic scale changes. The filters that perform this band division determine the multiresolution mother wavelets (PF-QMFs, etc.). The translation-scale (or time-frequency) resolution properties of this decomposition follow that shown in Figure 2.8. Carefully note that the center of each resolution cell corresponds to a point on the wavelet domain lattice. Thus, as was learned from the ambiguity function analysis, for some applications these resolution cells should all overlap at a point near the peak; otherwise, the wavelet domain representations become sensitive to noise (points on this surface well below the peak can be dominated by noise).

To satisfy the multiresolution properties, maintain noise immunity, and avoid aliasing, the multiresolution mother wavelets can become highly constrained. These constraints are worthwhile in a multitude of applications and are being advantageously exploited in image analysis. However, some of the properties of the multiresolution wavelet transforms may be undesirable for other applications.

As an example of an undesirable property, consider the decomposition of a linear FM signal. The wavelet transform of this FM signal can be superimposed on the multiresolution resolution cells that created Figure 2.8. Since the wavelet transform of a linear FM is a sloped line on the scale-translation grid, many coefficients (of approximately the same magnitude or energy) may be required to represent a majority of the FM's energy. A possible improvement for the compression application becomes obvious. If the mother wavelet was designed so that its resolution properties more closely matched the FM's characteristics as in Figure 3.3, then a more efficient representation (less coefficients to represent the same signal energy) could be possible. A linear FM time-scale representation overlaid onto Figure 3.3 might only overlap a few of the resolution cells; therefore, the energy of this FM signal is represented with only a few wavelet coefficients with respect to a FM mother wavelet. Under more uncertain conditions, multiple mother wavelets could possibly be employed to create an efficient representation. The mother wavelet design and multiple mother wavelet representations are detailed later, but are initially presented here to motivate the unconstrained wavelet analysis.

For multiresolution wavelet transforms the scaled and translated versions of the mother wavelet must satisfy some version of the orthogonality property (i.e., biorthogonality). To say that one signal is orthogonal to another is a very strong condition and requires detailed knowledge of the entire signal. For that reason arbitrary scale values cannot be chosen. If one function is the mother wavelet, then the scale values determine the form of the other functions (which must be orthogonal). For any mother wavelet this condition cannot be met for all scales. The multiresolution theory concentrates on only allowing scale factors that are a power of 2. This constraint on the scale sets up a rigid set of functions that must be further constrained to enforce the orthogonality conditions. For general, practical signals in many applications it is unacceptable to only have scales that are powers of 2.

Sampling Grids

Besides choosing a "good" mother wavelet, a wavelet domain grid (set of lattice points - particular scales and translations) must be specified. Normally, the multiresolution/orthogonal grids are dyadic grids (no freedom to choose another scale). When the mother wavelets are unconstrained and better resolution is desired, another grid can be chosen. Q: How is a "good" wavelet domain grid chosen? A: By considering the analogy to the narrowband correlation processing or matched filtering in which a set of hypothesized delays and frequency shifts forms a grid of hypotheses across the delay-frequency plane. The density of this grid is controlled by the allowable losses and the resolution properties of the transmitted signal's (mother wavelet's) resolution properties. Extending this concept to wideband correlation processing, a grid of wavelet domain "hypotheses" can be chosen by examining the mother wavelet's resolution properties and allowable losses.

Matched filter theory is well established [Van]. Wavelet transforms were related to WBCAFs in Chapter 3. The WBCAF was interpreted as the wideband correlator's output surface discussed in Chapter 2 and several references [Alt, Van, Zio]. Each hypothesized scale and delay indicates a grid point on these surfaces. How is the density of these hypotheses chosen for matched filter implementations? The grid density is proportional to the resolution of the signal. If the signal has very fine resolution in delay, then the hypotheses in delay are very close together or dense. If scale is highly resolved, then the scale lattice becomes dense.

The resolution of the signals is determined by the signal's auto ambiguity function. Examples have been shown in Figure 2.8 and Figure 3.3. These resolution cells are translated and scaled versions of the wideband *auto* ambiguity function (WBAAF) of the mother wavelet or the wavelet transform of scaled and translated versions of the mother wavelet:

$$\boldsymbol{WBAAF}_{a',b'}(a,b) \quad or \quad \boldsymbol{W}_g\left[\frac{1}{\sqrt{a'}}\,g\left(\frac{t-b'}{a'}\right)\right] \tag{4.1}$$

The density of the scale-translation evaluations, the (a',b') values, can be specified by the heights of the magnitude of this function. The magnitude of these WBAAFs can be placed across the scale-translation plane with one WBAAF at each (a',b') point. The density of the (a',b') evaluations depends upon the acceptable overlap of these WBAAFs. If the density is high, then the WBAAFs can be overlapped above their "1 dB down points," meaning that the ratio of the overlap point to the peak of the WBAAF is less than 1 dB. If a sparse density is sufficient, then the overlap of WBAAFs may occur at the 10 dB down points, etc. The extent of the overlap depends upon the specific application being considered and the degree of the tolerable distortion. These consideration have identical counterparts in narrowband theory (matched filter hypotheses densities) and these should be consulted for comparison in the wideband grid design. However, the grid density can be designed once a mother wavelet (any valid mother wavelet) is specified under the constraint discussed in Chapter 2, that states that the discrete scale step size must be approximately one.

Unconstrained Wavelet Transforms - Mother Wavelet Freedom

Now the book retreats to the general continuous time wavelet theory. The initial motivations for this step are to eliminate or reduce the previously mentioned constraints and to extend wavelet theory with a new operator that maps one wavelet transform to another. The mother wavelets are not required to be multiresolution or orthogonal and are, in general, "unconstrained." However, efficient representation can still be achieved by using multiple mother wavelets (or the best mother wavelet) rather than one constrained mother wavelet. This mother mapper operator efficiently utilizes multiple mother wavelets and is detailed and justified in the following sections.

As previously stated, most of the initial research in wavelet theory has concentrated on the orthogonal and biorthogonal wavelets. These wavelets have several desirable features that can be used advantageously, especially in image processing. However, the biorthogonal-orthogonal mother wavelets are created by specific filter structures under rigid constraints. These mother wavelets represent only a small portion of admissible mother wavelets. The remaining mother wavelets represent the "unconstrained" wavelet transform kernels. This section considers the utilization of these unconstrained wavelets.

One of the undesirable features of the biorthogonal-orthogonal wavelets is that their structure is constrained and thus their characteristics are not always controllable. The number of cycles and the rate of change of these cycles in a mother wavelet controls the resolution, compression and gain characteristics of the wavelet representation. Biorthogonal-orthogonal mother wavelets have fixed numbers of cycles and fixed rates of cycle changes; these numbers and rates may be altered by changing the order of the filters used to create the mother wavelets but full control over these parameters does not exist.

The primary action of these biorthogonal transforms is subbanding or band-splitting [Bur, Mal, Rio, Vai, Vet]. Subbanding operates on the spectral representation to achieve a multiresolution decomposition. Due to this repeated splitting action the fine resolution can be achieved even with biorthogonal wavelet filters that have only a small number of coefficients. Thus, only small order filters are constructed.

When the multiresolution/biorthogonal-orthogonal constraints are relaxed, allowing for unconstrained wavelets, the resolution characteristics can be easily altered by changing the mother wavelet. Specifically, the action of these unconstrained wavelet filters is not limited to band-splitting and will, in general, have a two-dimensional characteristic. Figure 2.8 displays the subbanding action in the time-frequency plane. Compare that subbanding representation (resolution cells that are vertical rectangular blocks) to the unconstrained wavelet (FM mother wavelet) represented by resolution cells in Figure 3.3. Note that the two-dimensional area is covered by resolution cells with a unique curvature. Other unique curvatures in the scale-translation space exist and can be achieved with different mother wavelets. Each different mother wavelet can provide an entirely different representation.

The justification for introducing ambiguity functions in the previous sections is that these time-scale representations are projections of the signal onto scaled and translated wideband auto ambiguity functions of the mother wavelet (the resolution cells in the figures). Understanding the ambiguity functions and their properties can be mapped to a comprehension of wavelet transforms and their resolution and representation properties.

By employing unconstrained mother wavelets, many desirable characteristics of the representation can be achieved. The "desirable" characteristics are application dependent. Later in this book several applications will be considered and the freedom in choosing a variety of properties (dependent upon the mother wavelets) will be used to advantage.

The resolution characteristics of the wavelet representations have been previously examined [Coi, Com, Dau, Vet]. The wideband cross ambiguity functions (WBCAFs) and wideband auto ambiguity functions (WBAAFs) are, under some conditions or constraints, also wavelet transforms.

Before detailing some of the advantages of exploiting unconstrained wavelet transforms, wavelet theory is extended by a new operator, the Mother Mapper Operator. The mother mapper operator leads to enhanced efficiencies for creating unconstrained wavelet transforms and is thus discussed before the applications of unconstrained wavelets.

The Mother Mapper Operator

The wavelet transform domain characterization of a signal and/or system is highly dependent upon the mother wavelet that is chosen as the kernel of the transform. A new operator, the **Mother Mapper Operator**, is derived to efficiently evaluate several transforms of the same function with respect to different mother wavelets. The Mother Mapper Operator **directly** maps a transform with respect to one mother wavelet to a new transform with respect to a new, different mother wavelet. The Mother Mapper Operator leads to signal/system representations which can involve **multiple mother wavelets** (or just the "best" mother wavelet) instead of just the original mother wavelet.

To efficiently evaluate wavelet transforms with respect to several different mother wavelets, a new operator, the Mother Mapper Operator, is presented. This operator will allow several mother wavelets to be considered without significantly increasing the required computational demands. The operator can also be applied to wideband cross ambiguity function (WBCAF) generation to enhance the efficiency and restructure of the WBCAF processor (Chapter 3).

Mother Mapper Operator Definition

The Mother Mapper Operator, $\mathbf{MM}_{g_2}(\cdot)$, maps $(\mathbf{R}\backslash\{0\} \times \mathbf{R}) \rightarrow (\mathbf{R}\backslash\{0\} \times \mathbf{R})$; i.e., ***it maps a wavelet transform of a function with respect one mother wavelet to a wavelet transform of that same function with respect to a different mother wavelet.*** The operator is defined as:

$$\mathbf{MM}_{g_2} : \mathbf{W}_g f(a,b) \mapsto \mathbf{W}_{g_2} f(a,b) :$$

$$\mathbf{W}_{g_2} f(a,b) = \frac{1}{c_g} \int\int \mathbf{W}_g f(a',b') \, \mathbf{W}_g^* g_2\left(\frac{a'}{a}, \left(\frac{b'-b}{a}\right)\right) \frac{db'\, da'}{(a')^2} \tag{4.2}$$

where $\mathbf{W}_g f(a,b)$ is the wavelet transform of some function, f, with respect to the mother wavelet, g. The positive real constant, c_g, is a normalizing term which is a function of the original mother wavelet only. This constant was defined previously as the admissibility constant.

The mother mapper operates as shown in Figure 4.2. For each point (a,b) in the output wavelet transform, the original wavelet transform (two-dimensional) is multiplied by another "warped" wavelet transform of the new mother wavelet, and the resulting two-dimensional surface of values is summed together to create one coefficient of the output surface. When considered as a multidimensional operator, the mother mapper is a four-dimensional operator that maps a two-dimensional surface to a new two-dimensional surface. Note that the surfaces shown in the Figure 4.2 are only *real* surfaces; in general, these surfaces will be complex, requiring even greater dimensionality. This example is only a representative illustration and the surfaces were not generated from specific signals.

The kernel of the Mother Mapper Operator is thus the wavelet transform of the new mother wavelet, g_2, with respect to the old mother wavelet, g. In this operator, the kernel is conjugated and then mapped via an affine transformation over all (a',b') (in two dimensions). See Figure 4.3. Next, these affine mapped kernels are functionally (or term-by-term in the discrete case) multiplied by the original wavelet transform and then integrated. Since the kernel of this new operator is a wavelet transform of the new mother wavelet with respect to the old mother wavelet, it is possible that it will have approximately compact support (or some type of structured or limited support) that may lead to enhanced efficiencies in implementations.

The limited significant support of this kernel is expected because the mother wavelets only have support near zero time (or space) and also have good scale-translation localization properties. Thus, a wavelet transform of a mother wavelet with respect to another mother wavelet will most likely have significant support in only a small region of the scale-translation plane.

After the mother mapper operator is derived, it will be used to rederive the wavelet domain WBCAF generator. Other applications have already been discussed and still others will be addressed in Chapters 5 and 6. The mother mapper operator can be applied any time multiple mother wavelets are being considered.

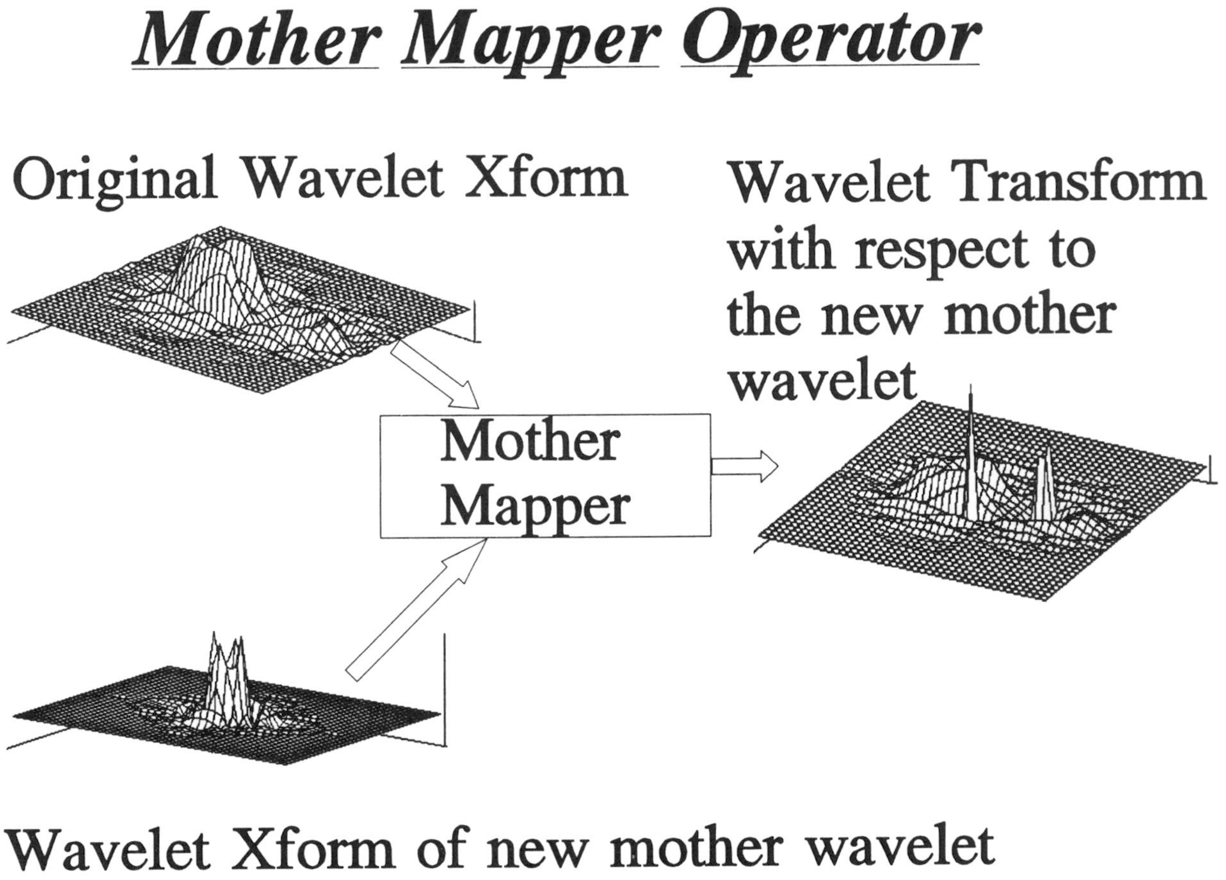

Figure 4.2: Mother Mapper Operation

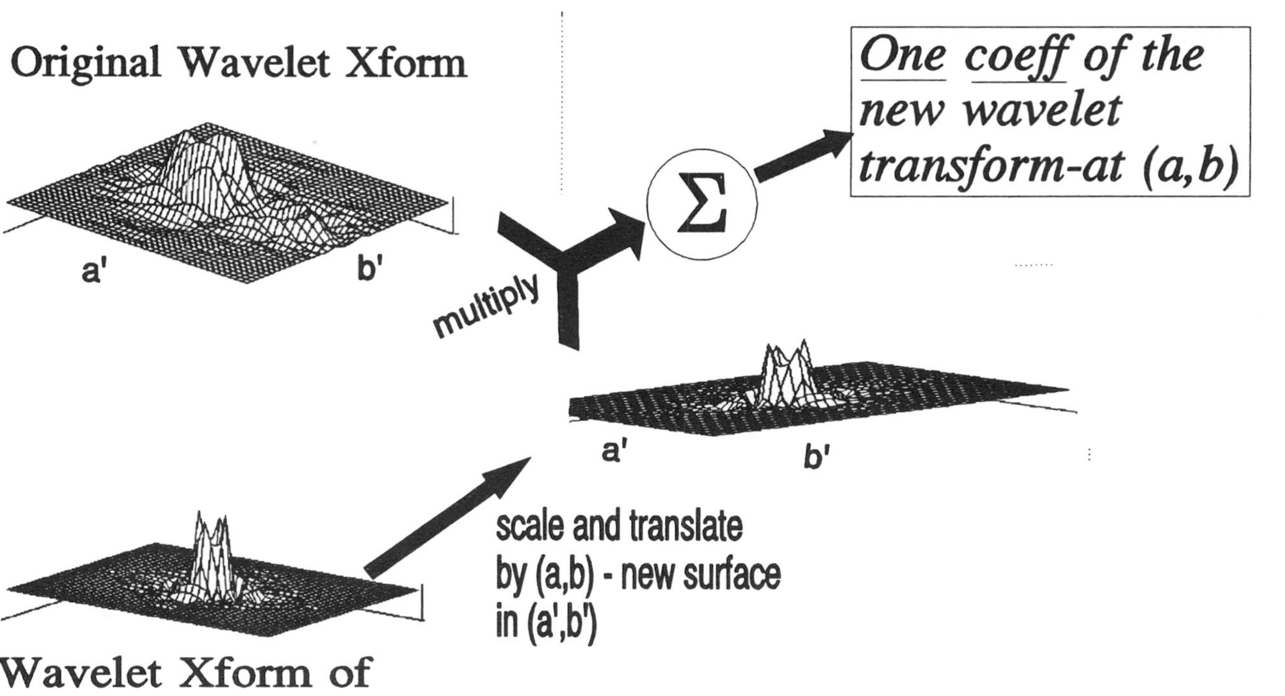

Figure 4.3: Two 2-D Surfaces yield One MMO Point

Mother Mapper Operator Derivation

The *resolution of identity theorem* (see Chapter 2) states that when g is an admissible function then:

$$c_g\langle f, f_2\rangle = \int_{-\infty}^{\infty} \frac{1}{(a')^2} \int_{-\infty}^{\infty} \langle f, g_{a',b'}\rangle\langle g_{a',b'}, f_2\rangle \, db' \, da' \tag{4.3}$$

The admissibility constant, c_g, must be a well defined constant (less than infinity). Let f_2 be a function of two new variables, a, b, becoming $f_2 = g_{2,a,b} = \frac{1}{\sqrt{a}} g_2\left(\frac{t-b}{a}\right)$. Under the assumption that g_2 is an admissible function, formulate the new wavelet transform (of f with respect to the new mother wavelet g_2) as an inner product:

$$W_{g_2} f(a, b) = \langle f, g_{2_{a,b}}\rangle \tag{4.4}$$

where a is the new scale variable and b is the new translation variable. Thus rewriting the resolution of identity yields:

$$\begin{aligned} W_{g_2} f(a,b) &= \langle f, g_{2,a,b}\rangle \\ &= \frac{1}{c_g} \int_{-\infty}^{\infty}\int_{-\infty}^{\infty} \langle f, g_{a',b'}\rangle \langle g_{2,a,b}, g_{a',b'}\rangle^* \frac{db' \, da'}{(a')^2} \end{aligned} \tag{4.5}$$

The second inner product on the right hand side of this equation is $W_g g_{2_{s,\tau}}(a,b)$ and it can be expanded as:

$$\begin{aligned} &\langle g_{2,a,b}, g_{a',b'}\rangle \\ &= \left\langle \frac{1}{\sqrt{a}} g_2\left(\frac{t-b}{a}\right), \frac{1}{\sqrt{a'}} g\left(\frac{t-b'}{a'}\right)\right\rangle \end{aligned} \tag{4.6}$$

By a transformation of the time variable, $t' = \frac{t-b}{a}$:

$$\begin{aligned} &\langle g_{2,a,b}, g_{a',b'}\rangle = \\ &= \left\langle g_2(t'), \frac{1}{\sqrt{\frac{a'}{a}}} g\left(\frac{t' - \left(\frac{b'-b}{a}\right)}{\frac{a'}{a}}\right)\right\rangle = W_g g_2\left(\frac{a'}{a}, \frac{b'-b}{a}\right) \end{aligned} \tag{4.7}$$

Returning to the resolution of identity formula with the above reformulation of the second inner product term yields the **Mother Mapper Operator:**

$$W_{g_2} f(a,b) = \frac{1}{c_g} \iint W_g f(a',b')\; W_g^* g_2\left(\frac{a'}{a}, \left(\frac{b'-b}{a}\right)\right) \frac{db'\,da'}{(a')^2}$$

$$(4.8)$$

Thus, the Mother Mapper Operator maps a wavelet transform of f with respect to g to a new wavelet transform of f with respect to mother wavelet, g_2. This operation is presented pictorially in Figure 4.2.

As a special case, let the mother wavelets be equal, $g_2(t) = g(t)$. The "reproducing kernel" property is produced.

$$W_g f(a,b) =$$

$$\frac{1}{c_g} \iint W_g f(a',b')\; W_g^* g\left(\frac{a'}{a}, \left(\frac{b'-b}{a}\right)\right) \frac{db'\,da'}{(a')^2} \qquad (4.9)$$

$$= \frac{1}{c_g} \iint W_g f(a',b')\; WBAAF_g\left(\frac{a'}{a}, \left(\frac{b'-b}{a}\right)\right) \frac{db'\,da'}{(a')^2}$$

This property states that the resolution of a wavelet transform can only be the resolution of its wideband auto ambiguity function or its reproducing kernel [Com]. Thus, the wideband auto ambiguity function of the mother wavelet is a good measure of the best resolution properties that a wavelet transform can achieve. This was one of the primary justifications for inserting ambiguity analysis in this book. Ambiguity and uncertainty functions have been applied and studied by a wide range of researchers and many of their general tradeoffs and properties are well appreciated for specific applications. This formulation also justifies the desired freedom in choosing a mother wavelet - the mother wavelet dictates the characteristics of the transform.

The mother mapper operator improves the efficiency of the unconstrained wavelet transforms. Unconstrained wavelet transforms and the mother mapper operator are expected to be utilized together. The remainder of this chapter concentrates on the unconstrained wavelet transforms.

Unconstrained Wavelet Transforms/Mother Mapper Operator Properties and Applications

The unconstrained wavelet representations include all of the information of the original function (signal or system operation). If a particular application determines that wavelet theory is required, desirable, or should be evaluated, then, at some point, a wavelet transform is formed. This initial wavelet transform has properties related to its chosen mother wavelet. However, a "better" representation may be achieved by choosing an alternative mother wavelet (an example of representing an FM signal was presented in Chapter 3 regarding a comparison of the unconstrained wavelet transform to an orthogonal multiresolution wavelet transform). Evaluating several different mother wavelets with significantly different characteristics will most likely lead to requiring some of these to be unconstrained wavelets.

Consider a speech signal. The vowel formants are relatively long duration tonal signals. The consonant-vowel and vowel-consonant transitions appear to be linear or quadratic FM's that chirp up or down. The glottal closings are sharp transitions. Other transient signals exist. The duration and sharpness of each of these signal components changes for different phonetic sounds and different speakers. If a set of mother wavelets are chosen that match each of the signal "features," then these features can be extracted over multiple scales and analyzing intervals (windows). Although the multi-scale properties of the general wavelet transform produce a representation that covers the time-scale space with a constant-Q structure, a more efficient representation may be achievable by using mother wavelets with resolution cells that have different orientations and structures matched to the features of the original signal. The mother mapper operator will efficiently create such representations.

More generally, suppose a signal, $f(x)$, is represented by a wavelet transform with respect to a particular mother wavelet, $g(x)$, and that this transform domain representation has nearly compact support (most of the coefficients are zero and only a few, big coefficients represent the entire signal). Now suppose a second mother wavelet, $g_2(x)$, leads to a transform domain representation with broad, spread out support (noncompact with no dominant coefficients). Obviously, if the information in $f(x)$ is to be saved or transmitted, then the compact representation (provided by the mother wavelet $g(x)$) will lead to efficiencies. Extending this idea, many different mother wavelets can be applied. The peaks of the most concentrated wavelet transform domain representation should be transmitted (transmit the coefficients from the transform with the signal energy concentrated in the fewest number of wavelet coefficients along with the particular mother wavelets as well). See Figure 4.4. Thus many different mother wavelets may be used to create an efficient representation. See Figure 4.5. Besides this ad hoc algorithm, more optimum representation techniques exist and are subsequently presented. However, the primary result is that multiple evaluations with other mother wavelets can lead to the energy of the original function being concentrated into a smaller set of coefficients.

Besides the applications of the mother mapper operator, the efficiency of the mother mapper operator is presented. One efficiency is that it avoids saving the original functional data or reconstructing that data from a previous wavelet transform. Figure 4.6 shows the concepts pictorially. Referring to the top row of Figure 4.6, first take a wavelet transform of the input signal (function) with respect to a initial mother wavelet; if the transform domain representation is "good enough" then stop and use that representation. If the representation is not good enough, then another mother wavelet might be used to decompose the signal (the idea is to proceed to the bottom right corner of Figure 4.6). At this point, without utilizing the Mother Mapper Operator, either the original data would have had to have been saved or that

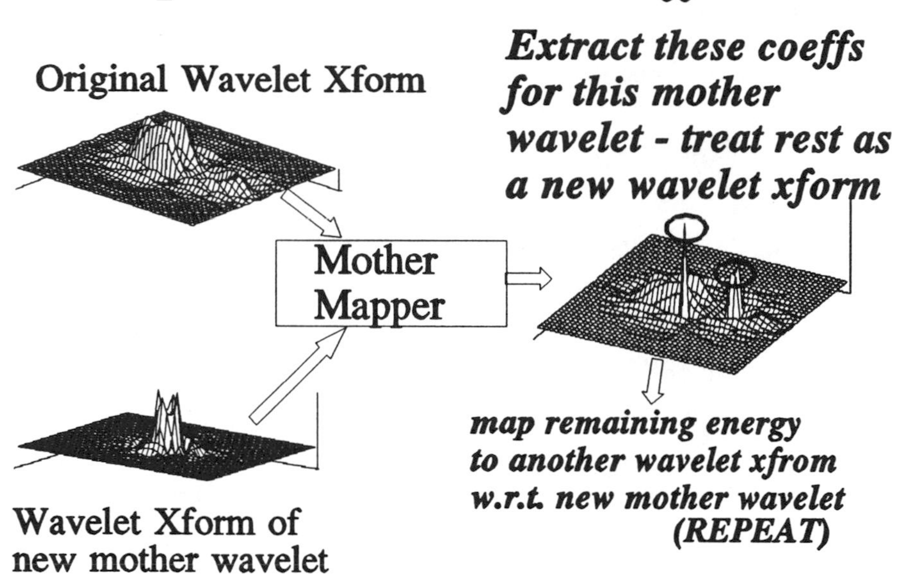

Figure 4.4: Multiple Mother Wavelet Efficiencies

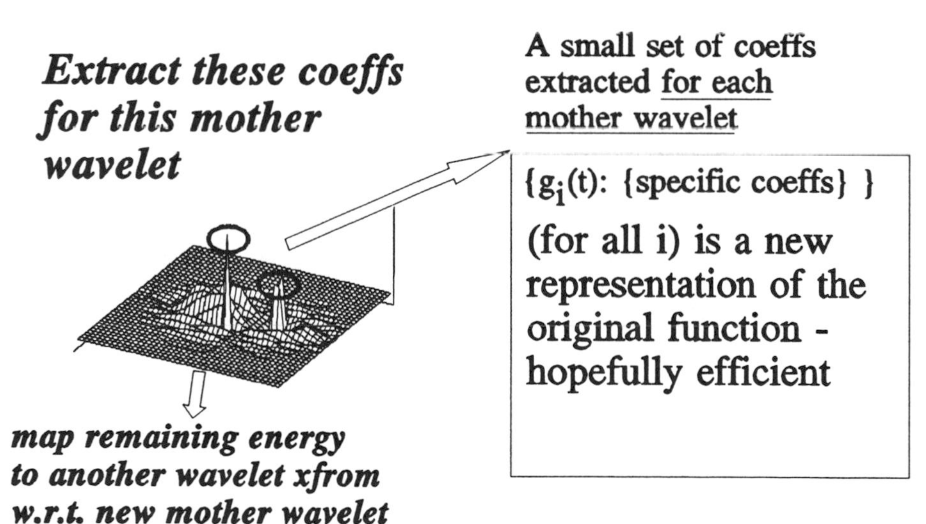

Figure 4.5: Multiple Mother Wavelet Representation

data would have to be reconstructed from the first wavelet transform (Figure 4.6). Saving the original data involves large resources as does reconstructing the data from the first transform. Instead, the Mother Mapper Operator provides a more efficient alternative by directly mapping the first wavelet transform to a new wavelet transform with respect to a different mother wavelet. Thus the efficiency of evaluating several transforms with respect to multiple mother wavelets is significantly improved.

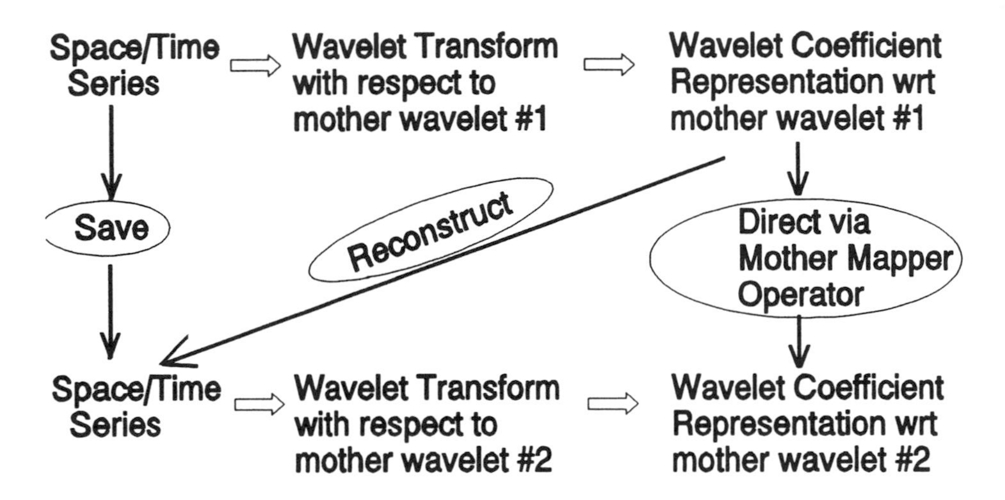

Figure 4.6: Mother Mapper Operator Efficiency

An advantage of using the Mother Mapper Operator to easily and efficiently map between transforms with different mother wavelets is that different parameterizations provided by different mother wavelets may produce efficient representations of the original process. Figure 2.3 and Figure 2.5 display wavelet transforms of the same rectangular signal but with respect to different mother wavelets. Figure 2.3 uses the Morlet, Gaussian weighted tone, as the mother wavelet while Figure 2.5 uses a highly modulated FM (with a Gaussian amplitude) as the mother wavelet. These figures emphasize the effects of the mother wavelet choice and how it parameterizes the wavelet transform.

Some processes will be "well represented" by a particular mother wavelet and other processes are poorly represented by that same mother wavelet. The concentration of energy in the transform domain represents the measure of efficiency or performance since a smaller number of coefficients would be required to represent the signal to the same degree of approximation.

One "trick" to efficient wavelet representations (a good approximation of the original signal with few wavelet domain coefficients) is to find the "optimum" mother wavelet or set of mother wavelets for the signals being analyzed (the optimality criteria are arbitrary and chosen by the user's applications). The orthogonal or biorthogonal wavelets are an example of picking good mother wavelets - these can be utilized to create extremely efficient dyadic wavelet transforms. Besides just picking a mother wavelet, these mother wavelets themselves may be optimized. Instead, if a set of mother wavelets is used, the set of mother wavelets should have different characteristics, such as their wideband auto ambiguity functions possessing different shapes and sizes. Further specific optimization criteria or algorithms are not addressed and are deferred to future research; however, the implementation of these algorithms would require efficient evaluations of wavelet transforms with respect to different mother wavelets. Another advantage of changing mother wavelets is that the mother wavelet parameterizes the transform in a manner similar to *parametric* spectral estimation. If the signals being transformed are similar to the mother wavelet, then the transform domain representation will tend to be efficient and characterize the salient properties of the signal or system. If one portion of a signal can be very efficiently characterized by a first, arbitrary, mother wavelet but another portion of the signal is inefficiently represented by this first mother wavelet, then the overall representation (the mother wavelet and its associated coefficients) will be inefficient. However, if a second mother wavelet is used that can efficiently characterize the originally poorly represented portion of the signal, then two (or multiple) mother wavelets together may be used to yield an extremely efficient representation. See Figure 4.7 for another diagram of this multiple mother wavelet decomposition process.

The new representations will involve sets containing the *significant* wavelet coefficients, with one set of coefficients associated to each mother wavelet. Effectively each mother wavelet will extract only that portion of the signal that it can efficiently represent, and it will allow the rest of the signal to be processed by another different mother wavelet transform. The set of mother wavelets and their associated coefficients (which are hopefully small in number) is the new representation of the signal. Thus a new decomposition, not only in terms of wavelet coefficients but also in terms of mother wavelets, may be defined.

Mother Mapper Operator Applications

As has been stated several times, multiple mother wavelet representations may lead to efficient signal representations. The efficient representation can be employed in communication systems. Instead of just subband coding [Mal, Vai, Vet] the input data (this is a wavelet transform with respect to one mother wavelet and then quantizing and coding the output of each filter bank), the input signal can be decomposed with respect to multiple mother wavelets. See Figure 4.5. These coefficients can then be quantized to approximate the original input signal. Now, only the set of coefficients with respect to multiple mother wavelets is transmitted - if this set is smaller and still approximates the same energy as in the subbanding case, then the multiple mother wavelet case should be evaluated. Since the multiple

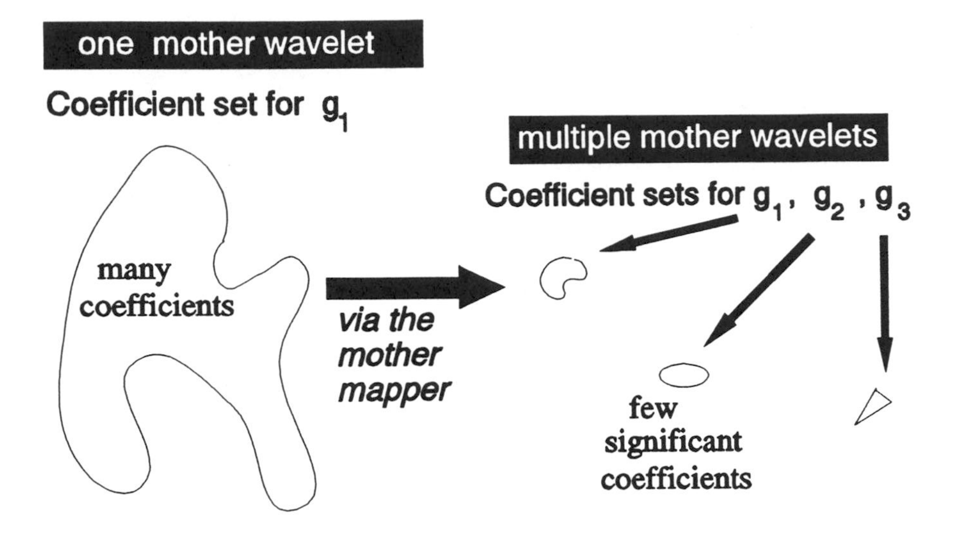

Figure 4.7: Multiple Mother Wavelet Representation

mother wavelet case is less efficient than the multiresolution case, the implementation must be considered. Since details of the implementation do not exist for the orthogonal case, this is again deferred for further evaluation. However, in the one-to-many broadcasting system the inefficiencies in the transmitter may be easily accepted.

Besides the mother mapper operator's utility for unconstrained wavelets, it can also be employed to reformulate established theories. Earlier, in Chapter 3, the WBCAF was reformulated in the wavelet transform domain. This reinterpretation provided insight by relating it to the structure of the NBCAF processor using FFTs in addition to increased efficiencies for some applications. By using the mother mapper operator, the derivation of the WBCAF reformulation in the wavelet transform domain becomes a simple substitution. Equations (3.29) provided the reformulation of the WBCAF, now by substituting $f(t)=r_1(t)$ and $g_2(t)=r_2(t)$ into the mother mapper operator, equation (4.2), and using equation (3.18) to create a WBCAF from a wavelet transform, the WBCAF is reformulated with much less effort.

Combining the ideas of WBCAF generation in the wavelet domain and efficient signal/system representations with multiple mother wavelets produces further efficiencies in WBCAF evaluations. By repeated application of the mother mapper operator, efficient signal representations may be created yielding a small set of coefficients with respect to each unconstrained mother wavelet. Then by substituting this few coefficient representation into the reformulated WBCAF, a wideband cross ambiguity function (WBCAF) can be more efficiently generated. By scaling the new representation (fewer coefficients) of one of the signals and then correlating it with the second received signal in the wavelet domain, a WBCAF is produced. Even if the correlation processing would take place in the time domain (or the domain of the appropriate independent variable) the multiple mother wavelet representation would lead to improved efficiencies.

Mother Mapper Operator - Final Considerations

Wavelet theory was extended to include a new operator, the Mother Mapper Operator. This operator allows sets of mother wavelets to be considered rather than just a single mother wavelet. The Mother Mapper Operator can improve the efficiency of implementing several wavelet transforms on the same data set with different mother wavelets (see Figure 4.4 and Figure 4.5). The Mother Mapper Operator allows a direct mapping from a previously computed wavelet transform to a new transform with respect to a new mother wavelet. The implementation is more efficient since the original space and/or time data does not need to be saved or reconstructed from the original wavelet transform in order to compute the new wavelet transform.

Unconstrained wavelet transforms are utilized rather than multiresolution orthogonal wavelet transforms to provide the freedom in choosing the mother wavelet rather than being highly constrained and yielding a very specific (and possibly unique) mother wavelet.

The mother mapper operator can improve the efficiency of WBCAF generation or cross wavelet transforms. These efficiency improvements are very general and, thus, can be achieved in many applications.

Earlier in this chapter, the cross wavelet transform was presented as well as its ability to extract the commonalties between two other wavelet transforms. That is the exact operation of the mother mapper operator; the wavelet transform of the new mother wavelet is compared to the wavelet transform with respect to the old mother wavelet. If portions of the old wavelet transform resemble the wavelet transform of the new mother wavelet, then those regions will yield a significant magnitude in the new wavelet transform. Thus, the commonality that is to be extracted (and indicated by a large magnitude) is the regions of the old transform that look like the transform of the new mother wavelet. For the mother mapper operation the scale-translation parameters of the output wavelet transform represent the same units as the scale-translation parameters of the input wavelet transforms. For the space-time processing, two temporal wavelet transforms can be cross wavelet transformed to map to a single wavelet transform in spatial coordinates (the axes are still scale and translation, but now they represent spatial coordinates). The form of the mother mapper operator is the same, regardless of the units.

Another justification for the mother mapper operator and the use of unconstrained wavelets is the space-time-varying system model presented next. The new system representation will exploit the mother mapper operator for efficient implementations and to form a wideband correlation model.

Further Research and Applications of the Mother Mapper Operator

By allowing general mother wavelets and creating more efficient implementations and mappings with those general wavelets, possibly adaptive or more robust mother wavelet schemes can be designed. The computational complexity of these new algorithms and efficient implementation considerations are critically needed. Due to the redundancy, parallel architectures should be considered.

Optimizing a set of mother wavelets for a specific application such as voice recognition or image analysis should be investigated. Now the orthogonality can be between different mother wavelets rather than a translated and scaled version of the same mother wavelet. Any comparison studies of the multiple mother wavelet techniques to other techniques should consider implementations as well.

Chapter 5: Linear Systems Modelling with Wavelet Theory

Introduction and Motivation

This chapter addresses system modelling with wavelet theory. The previous chapters presented the wavelet theory that is required for this chapter. Multiple, and possibly unconstrained, wavelet transforms and their resolution properties were accentuated. This freedom in the choice of the mother wavelet is critical for formulating system models with wavelet theory. This chapter constructs linear, time-varying system models with wavelet theory. This new system model exploits specific results of the previous chapters; it also justifies the topics covered in the previous chapters.

The further theoretical extensions in this book concentrate on formulating *system operations* and their applications with wavelet theory. It is important to note that the objectives of wideband-nonstationary *signal representations* are often very *different* than the objectives of time-space-varying *system representations*. Both the signal and the system representations exploit the wavelet theory features but the details of the resolution requirements are often significantly different; these differences and appropriate operators are detailed in this chapter. Wideband-nonstationary signals are often multiple octave signals that can be efficiently modelled with wavelet transforms, on a coarse scale grid, that are with respect to a particular mother wavelet (or set of mother wavelets). Space-time varying (STV) system models usually require a much finer resolution grid and are independent of any mother wavelet (the system is represented by a scale-translation distribution but the distribution is not relative to a particular mother wavelet).

As a specific example, the motion of a reflector (e.g., the motorcycle with rider) would have to be quite significant for it to scale a voice signal by a factor of 2 (e.g., a moving motorcycle might cause a scale of about 1.001); thus, system models should not, in general, be modelled by dyadic wavelets. Instead, fine resolution scale and translation representations must be constructed, justifying the previous chapters. This system model uses fine resolution mother wavelets but remains completely general (it can act on dyadic wavelet transform representations). Relative to the mother wavelet, the new system model formulates the mother wavelet as the input signal; therefore, the freedom in choice of the mother wavelet is required for the system to be an accurate model for many input signals.

Besides the resolution considerations, the new system model has a new operation that is wavelet related and is not the standard convolution. Some researchers have accepted convolution as the system operator even when systems are time-varying. Often, when wideband-nonstationary signals are involved the narrowband-stationarity conditions (required for the convolutional model) cannot be satisfied due to motion in the system. For those situations the system must be considered as time-varying over the duration and bandwidth of the signal. The convolution operator must be replaced by a more general operator. This time-varying system operator, the space-time-varying (STV) wavelet operator, has the same structure as an inverse wavelet transform.

For modelling systems both an operator and a system parameterization are required. The operator is fixed for each different system (i.e., convolution for linear time invariant systems) and the system parameterization changes for each different system (i.e., the impulse response for linear time invariant systems). The system parameterizations are usually required to represent the energy distribution in the system as well. The new model has a two dimensional system parameterization, the wideband system characterization. The wideband system characterization represents the distribution of the system's energy across scale and translation.

The new system model is consistent with the generally accepted wavelet transform definition that was provided in equation (2.2). This chapter uses wavelet theory to produce a new model for characterizing linear, space and/or time varying (STV) systems or channels. *This new STV system model includes the linear, time invariant (LTI) system model and the time-varying impulse response model as special cases*. Wavelet theory, which provides a better representation for some signals, can also provide a better characterization for some systems and characterize these systems more efficiently, more robustly, and with higher gain and/or resolution.

A linear time invariant system model creates the output signal from a weighted sum of delayed/advanced versions of the input signal. The STV wavelet operator models the output signal as a weighted sum of *scaled and* delayed/advanced versions of the input signal. *The time scaling allows the system model to track the variations of the system.* For some time-varying system models, such as a time-varying impulse response, the model estimation is performed over a time interval in which the system is assumed not to change (to be time invariant). These models include time and only one other parameter (a delay or frequency parameter) so that when estimating over a time interval, only a one-dimensional function is estimated (over delay or frequency). The STV model includes time and two other parameters, scale *and* delay. Thus, the STV model estimation creates a time-varying model at each instant of time and accounts for the system time variation over the estimation interval. *By properly tracking the variations in the system, longer processing intervals can be achieved in the system estimators.* The longer processing durations lead to more energy, higher gains, better resolution, and more robust modelling.

Prior to presenting the new system representation, previous, established system models are presented and formulated as parameterized operators; this same framework is later extended to the wavelet theory representation. To motivate the wavelet theoretic model, a practical example is provided in which the non-wavelet system models are sensitive and provide poor performance. The wideband, space-time-varying reflection process is the process to be modelled.

Wideband, space-time-varying (STV) signals and systems (channels) arise in many applications. The space-time-varying properties are sometimes referred to as nonstationary properties for statistical analysis. The space and/or time variation simply means that the system or channel evolves over space and/or time. For spatial systems space-varying could mean an *inhomogeneous* medium, such as the human body in which the properties are different at different spatial points in the body. However, this book concentrates on the more commonly addressed time variations exclusively (obviously space can replace time in any of the equations). In the non-non-wavelet system models, the time variation is often circumvented by processing over short time or spatial intervals during which the signals/systems are assumed to be approximately unchanged (invariant or stationary). The sacrifices that result from reducing the processing duration motivates the use of a better model.

Two new system models are presented. Both models are parameterized operators just like the established system models. The difference between the two new models is that one parameterization is independent of time (the parameterization is independent of time, but the operator is time dependent) while the other parameterization depends explicitly on time. Although the first of the new operators has a parameterization that is independent of time, this model is capable of representing systems that are time-varying. This apparent inconsistency is easily resolved when it is considered that the new *operator* includes time scaling (a time-varying operation). The second new model allows the parameterization to also be a function of time, analogous to the time-varying impulse response (which is a special case of the new model). By allowing the parameterization to be time-varying, then much more general systems can be modelled that include accelerations and higher order time variability.

The wavelet theoretic system model is exploited to model reflecting, absorbing, emitting and/or refracting sources or boundaries. The requirements for modelling such phenomenon arise in biomedical imaging, structural analysis, air and water acoustics, chemical reactions, and many other applications. The new system model should provide these applications with a better approach to characterize these processes and better techniques for forming estimates of these models.

Wideband/Nonstationary/Time-varying System Modelling

Since time-varying systems evolve with time, the time-varying system models must also evolve with time. If the system is observed for only a short interval, then, over that interval the system can appear to be time-invariant or slowly time-varying. Since the system, channel, or environment cannot be estimated with an infinitely long observation interval, a short (finite) observation interval is used. Most current estimators exploit this short observation interval to state that the system does not change over the observation interval (i.e., that the system is time invariant over this interval). The new system model accounts for some of the variation over the observation interval and attempts to model the change even over this short estimation interval - the new system model never assumes that the system is time invariant over any interval of time.

When the system changes over time but the system model does *not* change over time, the performance of the model degrades. How and why does the performance degrade? The performance degrades because the system is different over the *observation, probing, or processing interval.* The observation interval is the relatively short time window over which the system is modelled or estimated. This observation interval is chosen by the system modeler or designer and many tradeoffs must be considered before choosing a good observation or processing interval. Although this interval can be arbitrarily chosen to be very long, the system model may become *invalid* over this interval. When the system model becomes invalid, the gain(s), resolutions(s), and noise sensitivity of the system model or estimator degrade. The gains and resolutions of these system estimates are dictated by the *valid processing interval* which is the duration over which the system model can accurately characterize the system.

The valid processing interval depends upon the model chosen to characterize the system. If the system model is chosen to be time invariant (over the short interval), then the valid processing interval depends upon the system variation (velocities, acceleration, etc.) and the input/output signal's variability or characteristics (bandwidths, durations, etc.).

Consider the simple system that is composed of one moving reflector in the environment. Let the system input be the transmitted signal and the system output be the received (reflected) signal. The system's time variability is due to the motion of the reflector and the signal's variability is due to its bandwidth. For the narrowband system model (which assumes that the system is time invariant over the duration of the transmitted signal) the valid processing duration, T_p, is bounded by:

$$T_p < \frac{wavefront\_speed}{2\cdot(bandwidth)\cdot(reflector\_speed)} = \frac{c}{2Bv} \qquad (5.1)$$

which is a reformulated version of equation (3.10). This valid processing interval bound depends on both the signal variability (bandwidth, B) and the system variability (the reflector's speed, v) - the notation "v" is used to be consistent with related literature that uses "velocity" but here, velocity is not correct and it is replaced by the reflector's speed). The gain(s) and resolution(s) of the system estimate(s) generally improve as the valid processing interval increases. The estimation interval is typically chosen to be slightly longer than the transmitted signal duration. If a maximum reflector speed and a signal bandwidth are known, then a maximum valid processing interval can be established from equation (5.1). The signal to be transmitted is rarely chosen to be longer than the valid processing interval because no additional gain(s) or resolution improvement(s) can be achieved (and self interference may result due to phenomenon such as reverberation or clutter). Thus, the valid processing interval and the signal duration are usually the same. Essentially equation (5.1) states that the maximum motion in the system over the duration of the signal must be less than the resolvable motion. Equivalently, the system is not changing (relative to the observer's resolution) during the observation interval.

Equation (5.1) demonstrates that, for the narrowband or time-invariant system model, the valid processing interval decreases as the bandwidth and/or velocity (variability) increases. For electromagnetic environments the wavefront speed is the speed of light; for acoustic environments the wavefront speed is the speed of sound in a particular medium (e.g., air, water, materials, body, etc.). The velocity represents a closing velocity between the sources/reflectors and the sensors - the chosen reference frame is the sensor's reference frame and velocities are measured relative to the sensor (even if it is moving).

Consider the condition imposed by equation (5.1) for a voice signal in the air being emitted from a moving car to a listener that is standing near the road. Assume the car is moving at 10 meter/second (about 25 miles/hour). The speed of sound in air is about 325 meter/second. Thus, the relationship between the signal bandwidth and the valid processing interval is:

$$T_p \ll \frac{16.25}{bandwidth\ (Hz)} \quad seconds \tag{5.2}$$

For a voice signal with a minimal 2 kHz bandwidth, the valid processing interval, T_p, would be less than 1 millisecond or, stated differently, this time scaled signal can be accurately approximated by a Doppler shifted signal only over a very short interval. The Doppler shifting narrowband model would be valid over a longer period if the bandwidth were smaller; the standard physics model of a Doppler shift involves a train whistle (a narrowband tone) for which this model could be valid for as much as several seconds. With the voice signal and the short processing interval, negligible gain or resolution could be achieved by a narrowband processor. The narrowband or time-invariant model is not useful for the whole voice signal under these conditions (it could be valid for a short time at full bandwidth or for a longer time and only part of the bandwidth).

The new model accounts for some of the variability in the system and extends the valid processing duration to achieve higher gain(s) and finer resolution(s). *The primary improvement of the new model is the extension of the valid processing interval to achieve higher gains and better resolution.* Instead of the system model being limited by velocity (linear time scaling) in the real system, the limitation is now acceleration or quadratic time variation. With the new STV model the valid processing duration is limited by the bandwidth and the acceleration (velocities in the system can be modelled over any interval):

$$T_{p\text{-}WB} \ll \sqrt{\frac{c}{2 \cdot acc \cdot BW}} \tag{5.3}$$

This condition essentially states that the changes in the system's velocities (scales) must be small relative to the velocity (scale) resolution of the signal - only linear time variation is modelled (when it accelerates the model becomes invalid). Acceleration represents the rate of change of the velocity (scale). This condition is exactly analogous to the narrowband valid processing interval condition: the changes in position must be small relative to the position resolution of the signal. The two conditions are nearly the same if the narrowband conditions have position replaced by velocity and velocity replaced by acceleration. Before proceeding to detail the new system model, the general wideband reflection process is examined.

The Wideband Signal Reflection Process

A motivational application for the new model is the reflection of wideband signals from relatively fast moving objects. If a motionless "rigid" object reflects a signal, then the reflected signal is often modelled as being a delayed version of the incident signal. More generally, when the reflecting object is moving, the reflected signal can differ significantly from the incident signal. A particular reflection problem is presented in Figure 5.1. The sensor is both the transmitter and receiver. A simple time mapping due to the "head-on" collision of a constant velocity object and the travelling waveform (incident) signal is just a linear time scaling (see Figure 5.2). The reflected signal appears to be a time-warped and delayed version of the incident signal (without any acceleration terms); i.e. $r(t) = \frac{1}{\sqrt{2}} T\left(\frac{t}{2} - \frac{3}{2}\right)$ where $r(t)$ is the received (reflected) signal and $T(t)$ is the transmitted signal. A scale factor of 2 was arbitrarily chosen for simplicity (in reality no heart will move this fast but scaling a signal by 1.001 is very hard to recognize visually). This time-warping will be referred to as time scaling (scale) throughout the rest of the book. As discussed earlier in this book, time scaling is a feature of wavelet transforms.

Imaging Sensor - Heart Characteristics

Figure 5.1: Ultrasound Heart Imaging

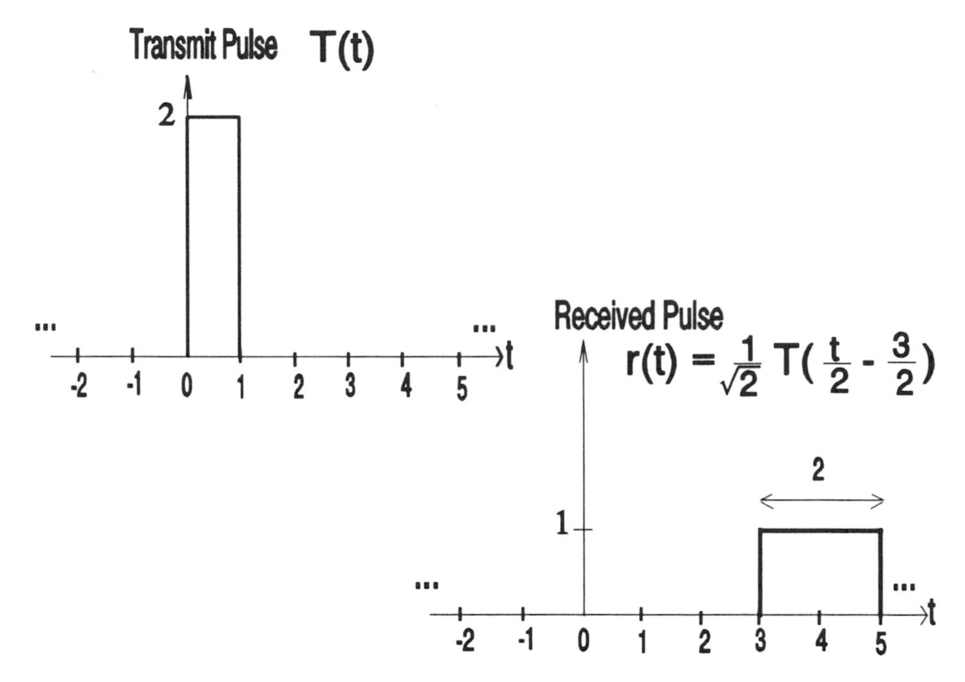

Figure 5.2: Output is Delayed and Scaled Input

A practical example of this reflection process on the travelling signal is now examined. Consider an ultrasound examination that noninvasively images a heart inside a body (see Figure 5.3 and Figure 5.1). The signal processing involves transmitting a signal, reflecting it off the heart, receiving the reflected signal, along with other corrupting signals, and processing the received signal to characterize the heart. For the purpose of simplification, the rest of the parts of the body are assumed not to affect the signal, and the reflected signal is assumed to be much stronger than any noise sources; thus, the only signal being considered is the reflection from the heart. Since the heart surface is assumed to be smooth and rounded, only one point is assumed to scatter (reflect) energy back into the receiver direction.

The main point of this example is to introduce motion of the reflector (the heart's pumping motion). *This motion must be non-negligible over the signal duration, meaning that the reflector must be at two different, distinct (resolvable) locations during the reflection process.* One location is when the front edge of the incident signal is being reflected and the second location is when the trailing edge of the signal is reflected. Consider the pumping action of the heart (or maybe even fibrillations of the heart surface) as shown in Figure 5.3.

For this example the reflecting surface is assumed to move away from the incident signal with a constant velocity. For numerical ease (and to exaggerate the effect) the surface is assumed to be moving half as fast as the incident signal wavefront (half the speed of sound in the body for this example). Figure 5.2 shows a possible transmit/receive signal pair. Note that the received signal is twice as long as the transmitted signal; the received signal is a scaled version of the transmitted signal.

The motion of the reflecting surface is depicted in Figure 5.5 which represents an elapsed interval of time over which the entire signal is reflected. Note that when the signal is first reflected, the heart is at its fullest expansion and closest to the receiver. As time progresses the heart contracts and the reflecting surface is moving away from the reflector. Finally a time is reached when the heart is fully contracted and the reflecting surface has moved its maximum amount away from the receiver. All of this motion occurs during the duration of the signal.

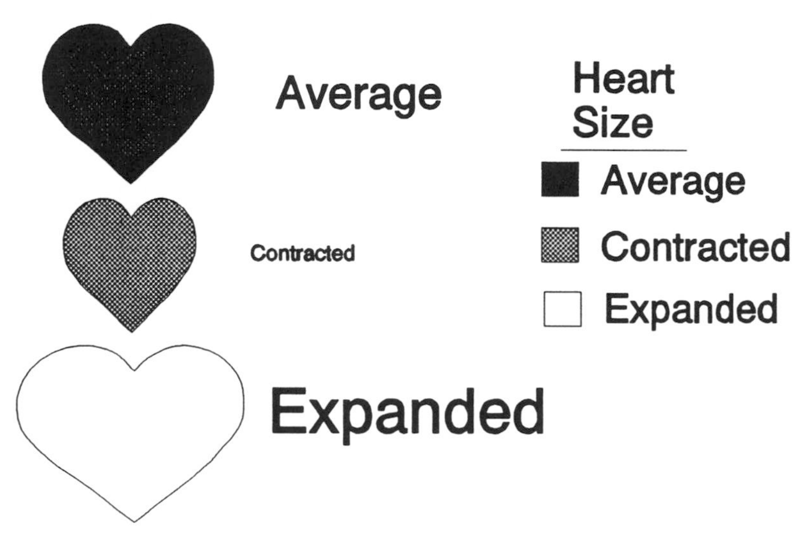

Figure 5.3: Heart Motion (Phases)

The time, t1, represents the time when the front edge of the signal is being reflected and the time, t3, represents the time that the trailing edge of the signal reflects.

The reflecting surface of the heart changes positions as the heart expands and contracts. If an incident signal has significant duration, then (see Figure 5.4) the location of the reflecting surface will change throughout the duration of the reflection.

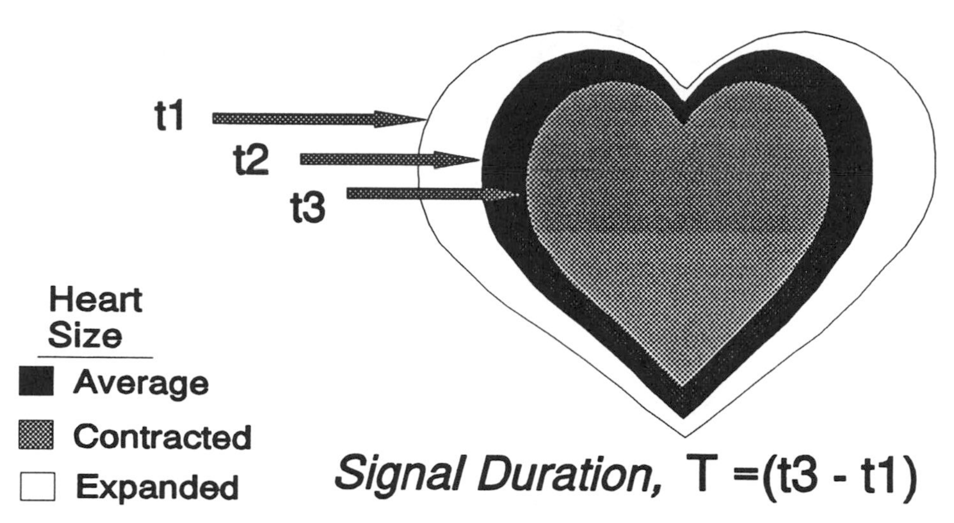

Figure 5.4: Reflector Position Changes over Signal Duration

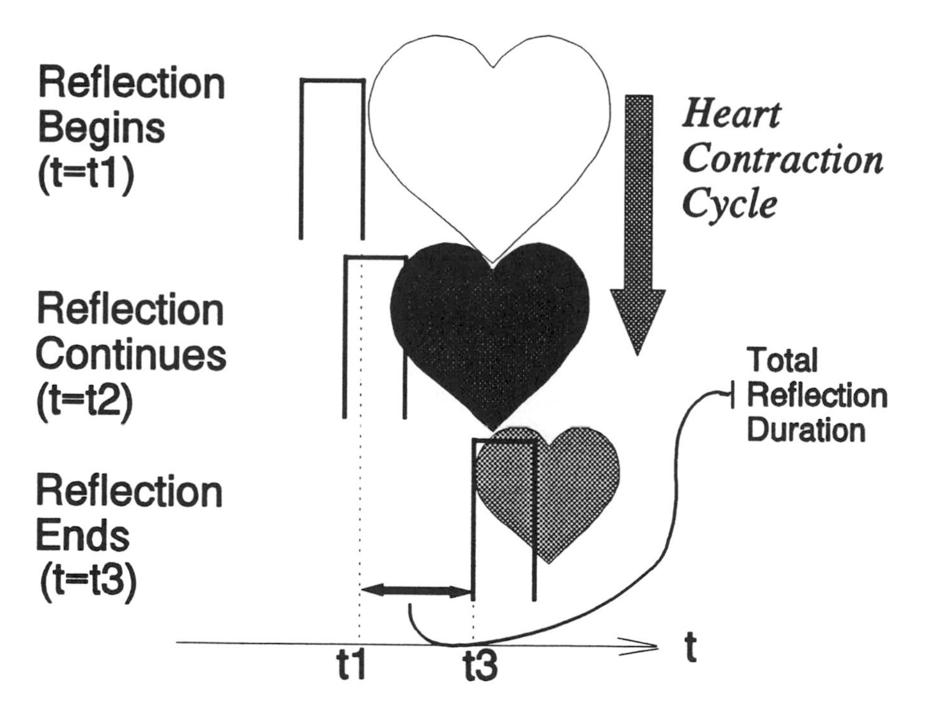

Figure 5.5: Signal Position During Reflection

Assume that the signal can resolve the different positions of the heart over this interval, then what is the model being estimated? Is the model a continuously changing image of the motion of the heart surface or is the model an estimate of the heart's position at one time instant? The non-wavelet models assume that the heart's position does not "resolvably" change over the signal duration. Thus, these models estimate one position over the duration of the signal (or some portion of the signal). *The new model accounts for the changing position of the heart surface over the signal interval and uses the entire signal interval to estimate its position at one snapshot in time.* By forming this estimate at different snapshots in time, a high gain and fine resolution image can be estimated at all times. This motivational example of the reflection process is continued with a specific transmitted signal being considered.

Wideband Reflection - Not Frequency Shifting

A particular wideband signal will be transmitted in the aforementioned, heart reflection example. Suppose the heart surface is moving at a constant rate of half the speed of sound away from the transmitter/receiver (see Figure 5.4). Assuming the transmitter is motionless, then, due to the heart's speed, the duration of the reflected signal is twice the duration of the transmitted signal. Thus the reflected signal is *time dilated* by a factor of two (to model the reflected signal, the time argument of the transmitted signal is divided by two).

Now suppose the signal consisted of just two simultaneous tonal pulses, one at 2 kHz and the other at 20 kHz (total bandwidth near 20 kHz). The reflected signal now consists of tonal pulses at 1 kHz and 10 kHz (total bandwidth near 10 kHz). The time scaling created a 1 kHz tone from the 2 kHz tone (a 1 kHz frequency shift for this component) and a 10 kHz tone from the 20 kHz tone (a 10 kHz shift for this component). *The reflected signal is not a frequency (Doppler) shifted version of the transmitted signal.* The frequency shift for each signal component can be significantly different when the system velocities and signal bandwidths are large. It is critical to note that the time-bandwidth product has not changed; the duration of the received signal is twice the duration of the transmitted signal but the bandwidth has been halved. Narrowband concepts are no longer valid in these situations and must be discarded. The wideband/wavelet concepts are required to adequately model this reflection process. In this example the *system model* can be better characterized by a wideband/wavelet model; wavelet theory applications are not limited to signal representations.

Consider the valid processing interval for the aforementioned example. Following equation (5.2), the narrowband system model is valid over a duration of approximately several milliseconds, and thus, the processing gain and resolution are also minimal. Now consider a new system model that can account for the true time scaling operation. The valid processing interval can be significantly extended. The new system model allows signal durations to be as long as the system's "non-accelerating" interval; the new system model is limited by the system's accelerations. When the heart transitions from contracting to expanding, an acceleration exists and the model fails. The valid processing interval for this new model is the duration during which the heart is consistently contracting (nearly linear motion); much longer than the valid processing duration of the narrowband system model.

If the signal and system conditions satisfy the narrowband conditions stated in equation (5.1) (see Chapter 3 for more details), then the time scaling action in the reflection process can be approximated by a Doppler frequency shift of the incident signal. The approximation fails as the signal becomes wideband or velocities increase so that the reflection process cannot be accurately modeled by a frequency shift. The new model will also fail when more general reflection is considered. The reflected signal can be more than just a time scaled version of the incident signal and the time-warping can have higher order terms due to the acceleration, and other higher order motions between the incident signal and the reflecting object. Since the wavelet transform only performs first order time scaling, even it alone is not adequate to model the higher order reflection process. Before considering these higher order terms, linear time scaling is modelled.

Consider the reflection example discussed previously. The reflected signal is a time scaled version of the transmitted signal. Stated with systems models: the output is a time scaled version of the input signal. Are present (non-wavelet) system modelling tools adequate to model this reflection process? Can wavelet theory be utilized to better model this reflection process? The detailed answers to these questions are answered only after the next section discusses present time-varying system models. The reflection example is referenced again and expanded upon after the system models are established.

Common Framework of System or Channel Characterizations:

Before the new system model is detailed, the previously established system or channel modelling techniques are presented and placed into a common framework. Previous system characterizations or models have always included some *operator*, $\boldsymbol{O}$, and a *parameterization*, $\{\boldsymbol{P}\}$. The parameterized operator maps the input signal to the output signal as shown in Figure 5.6.

Table 1 displays a set of operators and their associated parameterizations. The parameterization is termed the coefficient representation because the system model is typically estimated by a set of coefficients for all of these models. System identification is the estimation of the system parameterization or the coefficient representation. If the system is a human body, then the parameterization might be over delay and scale and, as discussed in Chapter 1, these parameters map to positions and velocities (of reflectors in the human body). Thus, the parameterization can be used to form an image of the body (this application is detailed in the next chapter). System and channel identification has been extensively investigated [Kai, Kuo, Pre, Zio, and many others]. For a chosen operator, each different system is characterized by a different parameterization. Besides the excitation, the response of the system is determined by both the system operator and its associated parameterization. One characterization for linear time-invariant (LTI) systems is the convolution operator and its associated parameterization is the system impulse response. All LTI systems may use the convolution operator but different LTI systems have different impulse responses (parameterizations). See Table 1.

Table 1: System Types, Operators, and appropriate Parameterizations

System Type	Operator	Parameterization
Linear, Time invariant (LTI)	Convolution with h(τ)	Impulse Response, h(τ)
Linear, Time invariant (LTI) Frequency Domain	$\boldsymbol{FT}^{-1}\,[H(\omega)\ \boldsymbol{FT}(input)]$	System frequency response, $H(\omega)$
Linear, Time-varying Superposition	$\int h(t,\tau)\,x(t-\tau)\,d\tau\,;\ x(t) = input$	time-varying impulse response, $h(t,\tau)$
Linear, Time-varying Frequency Domain	$\boldsymbol{FT}^{-1}\,[H(f,t)\ \boldsymbol{FT}(input)]$ $\boldsymbol{FT}(input)$ *is a function of f*	time-varying frequency response function, $H(f,t)$
Linear, Space and/or Time-Varying (STV) - LTI is a special case	$c_{input}\,\boldsymbol{WT}^{-1}_{(input)}\,[P(a,b)]$	Wideband system characterization, $P(a,b)$
Linear, Space and/or Time-Varying (STV) - (true time variation)	$c_{input}\,\boldsymbol{WT}^{-1}_{(input)}\,[P(a,b,t)]$	Time-varying wideband system characterization, $P(a,b,t)$

Note that ***FT*** and ***WT*** are Fourier and wavelet transforms, respectively and c_{input} is the admissibility constant of the input signal

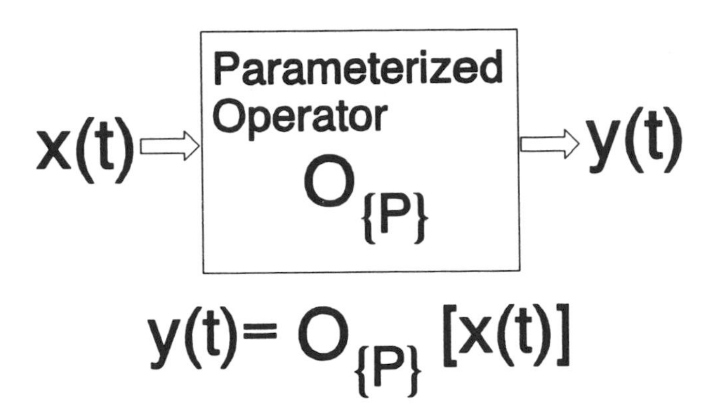

Figure 5.6: System Model Defined as an Operator

It is important to note that the same system may be modeled with different operators and parameterizations. The system action is represented by a parameterized operator, $O_{\{P\}}$, that acts on the input signal, x(t):

$$y(t) = O_{\{P\}}(x(t)) = Operator_{parameterization}(x(t)) \quad (5.4)$$

Several system characterizations exist and have been used. Depending upon the assumptions made regarding the system (e.g., linearity, time-invariance, etc.) several different system operators are valid. The assumptions regarding the system determine the class of characterizations that are valid. For example, LTI systems can be characterized by different system parameterizations (e.g. impulse response, step response, frequency response, etc.).

Assumptions regarding the system must be made before it is characterized. Several common assumptions are that the systems or channels are: deterministic or stochastic, linear, time-invariant, and/or stationary (if the system is modeled as stochastic). For stochastic systems the underlying probability distributions may also be assumed. Once the system assumptions are made, then one of the system operators from a class of systems operators that satisfy the assumptions is chosen.

System operators and *parameterizations* for the linear time-invariant systems and the linear time-varying systems are presented. For linear time-invariant (LTI) systems or channels one system operator is convolution, C, and the corresponding parameterization is the impulse response, $h(t)$. The corresponding input/output relationship is written as:

$$y(t) = C_{h(t)}(x(t)) = h(t) * x(t) = \int h(\tau)\, x(t-\tau)\, d\tau \quad (5.5)$$

For LTI systems a system is modelled by its impulse response. If the impulse response is not known but an input and output signal are known, then the impulse response can be estimated by operating on the input and output signals. For general systems this process of estimating the system's parameterization from input and output signals will be referred to as the *system (or system model) estimation problem.*

An equivalent LTI system characterization formulated in the frequency domain is a composite operator, $\mathcal{OF}$, made up of three other operators; the Fourier transform ($\mathcal{F}$), functional (term-by-term) multiplication, and the inverse Fourier transform ($\mathcal{F}^{-1}$). The parameterization is the system frequency response, $H(\omega)$. The input/output relationship is:

$$y(t) = (\mathcal{OF})_{H(\omega)}(x(t)) = \mathcal{F}^{-1}[H(\omega)\mathcal{F}(x(t))] \tag{5.6}$$

$$= \frac{1}{2\pi}\int H(\omega)\left[\int x(\alpha)\, e^{-j\omega\alpha}\, d\alpha\right] e^{j\omega t}\, d\omega \tag{5.7}$$

The step response and its convolution operator is another valid LTI model but it is not detailed.

Changing the assumptions regarding the system characteristics and requiring that the model handle time variation (and simultaneously maintaining the same structure as the LTI model) creates the next system model. If the system is linear but time-varying, then the system response may be characterized by the superposition operator, $\mathcal{S}$, and a time-varying impulse response, $h(t,\tau)$, (two-dimensional) [Zad, Zio]. The time-varying impulse response represents the system's time variation; at a particular time, t_o, the system responds with an impulse response, $h(t_o,\tau)$. The input/output relationship is:

$$y(t) = \mathcal{S}_{h(t,\tau)}(x(t)) = \int h(t,\tau)\, x(t-\tau)\, d\tau \tag{5.8}$$

Note that when the system is time invariant the time-varying impulse response becomes independent of time and this superposition becomes convolution.

Some researchers have transformed these time-varying characterization into the frequency domain for linear systems. The time-varying Fourier transform operator is $\mathcal{TVF}$ and the parameterization is Zadeh's time-varying frequency response function, H(f,t) [Zio]:

$$\begin{aligned} y(t) &= (\mathcal{TVF})_{H(f,t)}(x(t)) = \\ &= \int H(f,t)\left[\int x(\beta)\, e^{j2\pi f\beta}\, d\beta\right] e^{j2\pi ft}\, df \end{aligned} \tag{5.9}$$

with the relationship between the time-varying impulse response and the time-varying frequency response being related by a Fourier transform:

$$H(f,t) = \int h(t,\alpha)\, e^{-j2\pi f(t-\alpha)}\, d\alpha \tag{5.10}$$

For some nonstationary or time-varying systems these last two system models may be difficult to estimate. Using these models to represent first order time scaling may be difficult (estimating a system model of a system that includes time scaling can be sensitive and not robust). This is because these models only perform translation operations (across two time variables or across time and frequency, respectively).

Assume that the real system (that is to be modelled) performs time scaling; the output is a time scaled version of the input. This system must be modelled. Consider an input signal that has only one frequency. These previously presented models can effectively model the time scaling by performing a frequency shift of this single frequency. Now consider a frequency modulated (FM) signal that linearly sweeps its frequency. To properly time scale this signal, each different frequency component must be shifted by a different frequency. The system model in the frequency domain must map an input frequency component to the correct output frequency (but since the frequencies must be mapped instantaneously, the frequency does not really have much meaning). When the input signal has two or more different frequencies simultaneously, the time scaling operation is further complicated. The obvious problem is that the time scaling operation is not naturally modelled by frequency shifts when the signal has multiple frequency components as in the 2 kHz/20 kHz example. Wavelet transforms include the scaling operation, and thus, more appropriately model systems with linear time variation.

The properties of the existing time-varying system models appear to be undesirable for physical applications with "fast" moving reflectors and wideband signals. The reflected signal is a time scaled (possibly with linear and higher order time scalings) version of the incident wideband signal, not a frequency shifted version as assumed in some models. Different frequencies of the incident signal will incur different frequency shifts and are difficult to model. Typically, short processing durations are utilized to form a local or pseudo time invariant interval (even with these "time-varying" models).

Later it will be shown that the new STV wavelet system model includes the convolutional LTI system model and the time-varying impulse response models as special cases. From this previous discussion regarding the invalidity of frequency shifting, the frequency domain models may be sensitive if they have to be estimated. Thus, the new STV wavelet operator is more general and includes some of the previous LTI and time-varying models as special cases. The established system or channel models may be capable of modelling the time scaling operation but due to their forms, these models may be very difficult to estimate (estimating a continuous function of time by using some estimation interval must be carefully considered before it is blindly performed).

Is it important to model the time scaling or wideband reflection process? Yes, because by properly modelling this process the model will be valid over longer processing intervals; the longer processing intervals result in higher gains, more energy, better resolution, and more robust estimates. Reconsider the ultrasonic heart imaging example shown in Figure 5.1. With the standard narrowband theory, a valid model could be maintained only if the narrowband conditions on the signals bandwidth and duration of equation (5.1) were maintained. These requirements normally force the signal to have *either* good range resolution *or* good velocity resolution. By relaxing this stringent condition good measurements of both range and velocity can be achieved simultaneously. The simultaneous measurement of two variables is significantly more valuable than a sequential measurement of the two variables - the measurement is localized in a plane rather than at two points on two different lines (see [Che, Kna2] and Chapter 6).

The aforementioned system characterizations have been reformulated as parameterized operators. For each of the system models, the operator is always the same for all systems being characterized and only the parameterization is different for each different system (analogous to convolution being the only operator for LTI systems and each system being differentiated by its impulse response). A new system model defines both a new operator and a new parameterization. Now a new parameterized operator is defined for wideband-nonstationary signals and systems.

Note that some researchers have reformulated convolution with wavelet transforms [Dru1]. Although this research may provide numerical efficiencies for computing the convolution operation of LTI systems (possibly providing significant gains in image deblurring) or narrowband time-varying systems, it is not appropriate for wideband and/or nonstationary signals and time-space varying system or channel modelling because these systems will most likely not be time invariant (which is required for the convolution operator).

For LTI system models the duration of the valid processing interval dictates the available "processing gains." *These processing gains can simply be gains against noise sources or gains against other interfering sources, or the gain could be resolution improvements.* Typically, the longer the valid processing duration, the more efficient the processing becomes (larger block sizes can be processed). Since the invariant durations control the performance of the LTI systems the system designer attempts to maximize the invariant period. For some applications special observing sensor configurations can be engineered to accomplish extending the period; for other applications short periods, and thus low gains, must be accepted. Thus, a desirable property of a new system model is to extend the valid processing interval. *Typically, the larger the valid processing period, the higher the system gain. Therefore, modelling the system through some of its variation is highly desirable.*

A new system model that is capable of modelling the system through some of its variations is presented next. The new system model is termed the STV wavelet operator.

The STV Wavelet Operator - Space-time-varying System Model

This section introduces a new space and/or time varying system model, the STV wavelet operator. The characterization or modelling of space-time-varying (nonhomogeneous in space or nonstationary in time) systems should include the capability to robustly model or estimate the scaling operator. By employing this time scaling operation to account for the system's time variation, a system model *at one snapshot of time* can be estimated from a "long" time interval. Wavelet transforms employ the time scaling operator and will thus be utilized for modelling the space-time-varying (STV) systems. Wavelet transform theory and its advantages over narrowband approaches are discussed in the references [Com, Dau, Mal, Mey, Vet] and previous chapters and is not repeated here. Wavelet theory is not required to formulate the new operator but its close relationship justifies the name. Also, the properties of the new system model are easily established by considering wavelet theory. To avoid notational difficulties and provide consistency with the previous chapters, only the time parameter is used throughout the rest of this chapter. The spatial parameter can replace the time parameter and all of the mathematics will remain valid; however, the physical interpretations may be altered.

Strictly, the nonstationary condition refers only to stochastic signals and or systems; however, to avoid repeating that the mathematics are valid for both "nonstationary" and/or "time-varying" signals or systems, this chapter uses the nonstationary and time-varying properties interchangeably.

The new STV wavelet operator , $\boldsymbol{STV}_{P(a,b)}$, utilizes time scaling and maps the input, x(t), to the output, y(t), even when both the input signals and the system (modeled by $\boldsymbol{P}(a,b)$) are wideband and nonstationary. The **STV wavelet operator** is a composition of several operators and is defined as:

$$y(t) = \boldsymbol{STV}_{\boldsymbol{P}(a,b)}(x(t))$$

$$= \int_{-\infty}^{\infty}\int_{-\infty}^{\infty} \boldsymbol{P}(a,b)\ \frac{1}{\sqrt{a}}\ x\left(\frac{t-b}{a}\right)\frac{db\ da}{a^2} \qquad (5.11)$$

$$= c_x\ \boldsymbol{W}_x^{-1}[\boldsymbol{P}(a,b)]$$

where $\boldsymbol{W}_x^{-1}$ is an inverse wavelet transform with respect to the mother wavelet, $x(t)$ (which is the input signal) and c_x is the admissibility constant of the input, $x(t)$. The admissibility constant, c_x, is defined as:

$$c_x = \int_{-\infty}^{\infty} \frac{|\boldsymbol{X}(\omega)|^2}{\omega}\, d\omega \qquad (5.12)$$

where $\boldsymbol{X}(\omega)$ is the Fourier transform of $x(t)$ (the beginning of Chapter 2 discusses the admissibility constant in more detail).

Analogous to the previously presented system models, the system parameterization or coefficient representation is the **wideband system characterization,** $\boldsymbol{P}(a,b)$, and the system operator is the STV operator, $\boldsymbol{STV}$ (which is essentially an inverse wavelet transform). Refer back to Table 1 on page 155. For comparison purposes, the wideband system characterization, $\boldsymbol{P}(a,b)$, is analogous to the impulse response, $h(\tau)$, while the STV wavelet operator, $\boldsymbol{STV}$, is analogous to the convolution operator, $\boldsymbol{C}$. Clearly, the $\boldsymbol{STV}_{P(a,b)}$ operator is independent of x(t); it simply operates on x(t) as is required for a valid system model.

Physically, the STV wavelet operator creates an output by summing weighted, *scaled* and translated replicas of the input, x(t), with the weights being the wideband system characterization, P(a,b). This is analogous to the LTI system models in which the output is a weighted sum of translated replicas of the input signal. Obviously, time scaling is the additional feature of the STV wavelet representation and is also the key to efficient representations of the wideband reflection or scattering process and improved gains.

Besides the more easily analyzed STV wavelet operator of equation (5.11), a more general time-varying STV operator includes the time parameter in the wideband system characterization. This form was presented in Table 1 and simply adds a time variable to the wideband system characterization:

$$y(t) = \mathbf{STV}_{P(a,b,t)}(x(t))$$

$$= \int_{-\infty}^{\infty}\int_{-\infty}^{\infty} P(a,b,t)\,\frac{1}{\sqrt{a}}\,x\!\left(\frac{t-b}{a}\right)\frac{db\,da}{a^2} \tag{5.13}$$

$$= c_x\,W_x^{-1}[P(a,b,t)]$$

The STV operator has not changed; only the system parameterization has become a function of time. Now the time-varying wideband system characterization, $P(a,b,t)$, is analogous to the time-varying impulse response, $h(t,\tau)$. Later the time-varying impulse response is shown to be a special case of the time-varying wideband system characterization. In the next couple sections the time-varying wideband system characterization is not used and only the "time-invariant" or snapshot wideband system characterization, P(a,b), is employed to establish properties and analogies.

The time-varying wideband system characterization simply adds the time parameter to the wideband system characterization. Why is the original wideband system characterization (without the time parameter) called "time-varying" if its wideband system characterization does not depend on time? Consider the wideband system characterization to be a "time invariant" time-varying wideband system characterization or $P(a,b) = P(a,b,t_o)$ (the time-varying wideband system characterization evaluated at a "snapshot" of time). Although this system model is valid at this snapshot in time, it is also approximately valid for some interval around this snapshot instant of time. The wideband system characterization (P(a,b)) model is valid over intervals in which the system varies linearly with time; when accelerations exist in the system, the model becomes invalid. However, *this "time-invariant" time-varying wideband system characterization, P(a,b), is valid even over intervals in which the system varies (albeit linear with time).* Carefully note that even though the system model is valid over some interval, the system characterization is only valid at a snapshot instant of time.

If a single point reflector is moving at a constant velocity toward a transmitter/receiver and modelled as a system (with the transmitted signal being the input and the received signal being the output), then the STV's wideband system characterization will be a single impulse at a particular scale (velocity) and translation (delay or range) - this example is detailed later but is initially presented here to clarify the distinction between an estimation interval and an estimate. Since the model of this single reflector is an impulse, then only one velocity and range value are indicated. But the range value is constantly changing with time. Normally an image or model of the environment is formed at some instant of time; thus, the wideband system characterization would map to such an image. However, since the system only involves constant velocity an extremely long interval can be used to estimate the system model (wideband system characterization) but the estimate is at a snapshot instant of time. Since constant velocity motion maps to a time-varying system, the system is considered to be time-varying. Now, due to the time scaling operation in the STV operator, the system model can exactly track this time variation. So the "time invariant" wideband system characterization is a model or image of a time-varying system at a snapshot instant of time (an estimate) but due to the time variation incorporated into the model, a long processing or estimation interval can be used. To create an image or model (wideband system characterization) at all instants of time then becomes the time-varying wideband system characterization.

Both the "time invariant" and the time-varying wideband system characterization are models of the system's energy distribution and justification for this interpretation is the topic of the next section.

STV Wavelet Operator's Energy Distribution

Reconsider equation (5.11). The form of differential term in the STV wavelet operator, $\frac{db\,da}{a^2}$, is required. This differential term causes the wideband system characterization to represent the *distribution of energy in the system.* Without this differential form the wideband system characterization would not represent an energy distribution. The energy in the wavelet domain should be the same as the energy in the time domain (an isometric transform); this is the only differential that will satisfy this condition. For a wavelet transform of a signal, x(t), with respect to a mother wavelet, g(t), the differential amount of energy in x(t) in a particular scale and translation area centered at (a,b) is $|\boldsymbol{W}_g x\,(a,b)|^2\ \frac{da\,db}{a^2}$ as discussed in Chapter 2 (please refer to Chapter 2 for further details). Thus, since the STV operator is also an inverse wavelet transform, the distribution of energy in the wideband system characterization is $|\boldsymbol{P}(a,b)|^2\ \frac{da\,db}{a^2}$. For the wavelet transform defined in equation (2.2), any other differential terms would necessarily imply that the distribution in scale-translation would not represent the energy distribution in the system. Just as in narrowband analysis or LTI system analysis, the representation of the system should be comparable to the representation of the signals. Since the signal's energy distribution is $|\boldsymbol{W}_g x\,(a,b)|^2\ \frac{da\,db}{a^2}$, then the system's energy distribution should be $|\boldsymbol{P}(a,b)|^2\ \frac{da\,db}{a^2}$ as well. Any other definition of the system operator would cause the signal's energy to be modelled differently than the system's energy.

This differential also holds for scattering characterizations (system models of the reflection process) as well. This differential term allows the wideband system characterization to be interpreted as a wavelet transform; the STV wavelet operator then becomes an inverse wavelet transform with a constant scalar in front of it. The admissibility constant in front of this inverse wavelet transform is also critical. This constant allows the wideband system characterization to be independent of the input, as any system characterization must be. Besides satisfying the required properties for a system model, the STV wavelet model has several desirable properties as well.

Estimation of the Wideband System Characterization

By formulating the new system model as an inverse wavelet transform in equation (5.11), the estimator for the wideband system characterization, $\boldsymbol{P}(a,b)$ (or $\boldsymbol{P}(a,b)$) becomes simple. Taking the forward wavelet transform of equation (5.11) *with respect to the input signal as the mother wavelet* yields the estimate:

$$\begin{aligned} \boldsymbol{W}_x\,[y(t)]\,(a,b) &= \boldsymbol{W}_x\,[\boldsymbol{STV_P}(x(t))]\,(a,b) \\ &= c_x \boldsymbol{W}_x\,\{\boldsymbol{W}_x^{-1}[\boldsymbol{P}]\}\,(a,b) = c_x \hat{\boldsymbol{P}}(a,b) \\ thus \quad \hat{\boldsymbol{P}}\,(a,b) &= \frac{1}{c_x}\,\boldsymbol{W}_x\,[y(t)]\,(a,b) \end{aligned} \tag{5.14}$$

Note that this is not an equality; it is one particular estimator. The justification for this equation not being an equality is due to the non-uniqueness property of the inverse wavelet transform discussed in Appendix 2-A. In addition, this is just one estimator. If the estimated wideband system characterization is used to estimate point reflectors in an environment, then other estimator can further process this initial estimate (i.e., pick the maximum of the magnitude of this initial estimate or "deconvolve" as in the narrowband models [JohB]). Equation (5.14) is developed further in Chapter 6 when it is formulated as the wideband correlation receiver's output. The primary property of this estimator that is investigated further is the valid processing or estimation interval over which this estimator is an accurate model. First other general properties of the STV wavelet operator are considered.

Properties of the STV Wavelet Operator

The STV wavelet operator linearly maps an input signal to an output signal. The output is a weighted sum of scaled and translated versions of the input signal. The attributes of this output are controlled by the weighting function, the wideband system characterization, $\boldsymbol{P}\,(a,b)$. This section demonstrates that the LTI system model is a special case of the STV wavelet operator when the support of $\boldsymbol{P}\,(a,b)$ is limited to the line, a=1 (unity scale). This section also provides an example of the nonuniqueness of the system representation, $\boldsymbol{P}\,(a,b)$, and how the same output signal can be created from multiple $\boldsymbol{P}\,(a,b)$'s. Finally, more general wideband system characterizations are considered. Examples of systems with support in large regions of the scale-translation plane are presented.

Linearity

If an input signal, $x_{tot}(t)$, is formed as a weighted sum of two other inputs:

$$x_{tot}(t) = c_1 x_1(t) + c_2 x_2(t) \qquad (5.15)$$

and the output for each individual input is computed:

$$y_i(t) = \iint \boldsymbol{P}(a,b) \frac{1}{\sqrt{a}} x_i\left(\frac{t-b}{a}\right) \frac{db\,da}{a^2} \quad for\ i=1,2$$

$$y_{tot}(t) = \iint \boldsymbol{P}(a,b) \frac{1}{\sqrt{a}} x_{tot}\left(\frac{t-b}{a}\right) \frac{db\,da}{a^2}$$

$$= \iint \boldsymbol{P}(a,b) \frac{1}{\sqrt{a}} \left[c_1 x_1\left(\frac{t-b}{a}\right) + c_2 x_2\left(\frac{t-b}{a}\right)\right] \frac{db\,da}{a^2} \qquad (5.16)$$

and the system response to the weighted sum of two inputs becomes the weighted sum of the corresponding individual outputs:

$$y_{tot}(t) = c_1 y_1(t) + c_2 y_2(t) \qquad (5.17)$$

thus, the *STV operator is linear.*

STV Representation of LTI Systems

A LTI system can be modelled by the STV wavelet operator. The representation of a LTI system provides insight for the interpretation of the wideband system characterization. If the wideband system characterization only has support along the unity scale line (a=1) and is zero everywhere else $\boldsymbol{P}(a,b) = \delta(a-1)\,h(b)$, then the STV operator with this characterization is equivalent to the LTI operator of convolution with the impulse response, h(t). The justification follows:

$$y(t) = \iint \delta(a-1)\, h(b) \frac{1}{\sqrt{a}} x\left(\frac{t-b}{a}\right) \frac{db\,da}{a^2} \qquad (5.18)$$

allowing the first integral to perform its sifting operation yields:

$$y(t) = \int h(b)\, x(t-b)\, db = \int h(\tau)\, x(t-\tau)\, d\tau \qquad (5.19)$$

which is the convolution operator, parameterized by the impulse response of the LTI system. Therefore, *the slice of the wideband system characterization at a unity scale value represents the impulse response of a LTI system.* In addition, the more concentrated the wideband system characterization is along the unity scale line, the more the system acts as if it were LTI - this will be useful for the system identification application. Essentially, the STV wavelet operator generalizes the LTI operator to a higher dimensionality. Fortunately, wavelet transforms are a convenient tool for characterizing this higher dimensional space. It is intuitively pleasing that when the dimensionality of the STV wavelet representation is collapsed it represents a LTI system.

Time-varying STV Representation of Time-varying Impulse Responses

The time-varying system model with superposition and a time-varying impulse response can be modelled as a special case of the time-varying STV wavelet operator. The representation of the time-varying impulse response provides insight for the interpretation of the time-varying wideband system characterization. If the time-varying wideband system characterization only has support along the unity scale line (a=1) and is zero everywhere else $P(a,b,t) = \delta(a-1)\, h(t,b)$, then the STV operator with this characterization is equivalent to the time-varying impulse response and superposition operator in equation (5.8). The justification follows:

$$y(t) = \iint \delta(a-1)\, h(t,b)\, \frac{1}{\sqrt{a}}\, x\left(\frac{t-b}{a}\right) \frac{db\, da}{a^2} \qquad (5.20)$$

allowing the first integral to perform its sifting operation yields:

$$y(t) = \int h(t,b)\, x(t-b)\, db = \int h(t,\tau)\, x(t-\tau)\, d\tau \qquad (5.21)$$

which is the superposition operator, parameterized by the time-varying impulse response of the system. Therefore, *the (multidimensional) slice of the time-varying wideband system characterization at a unity scale value represents the time-varying impulse response of the system.* The STV wavelet operator generalizes the time-varying impulse response model by adding another dimension. Fortunately, wavelet transforms are a convenient tool for characterizing this higher dimensional space. It is intuitively pleasing that when the dimensionality of the time-varying impulse response is collapsed it represents a time-varying impulse response system.

The time-varying impulse response and its frequency domain representation have been extensively employed in imaging systems. This new, more general model leads to more robust estimators and should improve those applications that already employ these time-varying models. No further properties will be considered for the time-varying wideband system characterization model; only the "snapshot" wideband system characterization will be analyzed.

Consider a time-varying system that simply time scales and delays the input signal to create an output signal, $y(t) = x(a(t-\tau))$. Now consider the impulse response of this system:

$x(t) = \delta(t)$ *and* $y(t) = \delta(a(t-\tau)) = \frac{1}{|a|}\,\delta(t-\tau)$. Thus, the impulse response is an attenuated impulse at the same location in time! One problem with this example is that the delta function is not a square integrable function, but if this is a difficulty, the these models and their estimation must be done very carefully... This book concentrates on creating a model that can be directly estimated even for time-varying signals and systems.

Identity System

For the STV wavelet operator the *identity system* is represented by a wideband system characterization that is an impulse function, $P(a,b) = \delta(a-1)\,\delta(b)$. The identity system's output is just equal to the input.

operator, and the fact that the identity system is a LTI system, the impulse response of the identity system must, of course, be an impulse. So, returning to equation (5.11), the wideband system characterization must be $P(a,b) = \delta(a-1)\,\delta(b)$ for the identity system.

$$y(t) = \mathbf{STV}_{P(a,b)}(x(t))$$
$$= \int_{-\infty}^{\infty}\int_{-\infty}^{\infty} \delta(a-1)\,\delta(b)\,\frac{1}{\sqrt{a}}\,x\left(\frac{t-b}{a}\right)\frac{db\,da}{a^2} \tag{5.22}$$
$$= x(t)$$

Single Scale and Delay System

If the input signal is $x(t)$ and the output is $y(t) = \frac{1}{\sqrt{s}}\,x\left[\frac{t-\tau}{s}\right]$ then the wideband system characterization is simply an impulse at (s,τ) in the scale-translation plane; thus $P(a,b) = \delta\left(\frac{a}{s}-1\right)\delta\left(\frac{b-\tau}{s}\right) = s^2\,\delta(a-s)\,\delta(b-\tau)$. Substituting this characterization into the STV wavelet operator yields:

$$y(t) = \iint s^2\,\delta(a-s)\,\delta(b-\tau)\,\frac{1}{\sqrt{a}}\,x\left(\frac{t-b}{a}\right)\frac{db\,da}{a^2}$$
$$= \frac{1}{\sqrt{s}}\,x\left(\frac{t-\tau}{s}\right) \tag{5.23}$$

Thus, the wideband system characterization for a single point scatterer at a particular scale and delay is determined. Details of the "scattering" interpretation are deferred to Chapter 6. Any input will lead to an output that is a scaled and delayed version of the input. Note that the output signal has the same energy as the input signal for this particular wideband system characterization; the wideband system characterization requires the square of the scale term to normalize the energy.

Can this time scaling be used to approximate frequency shifting for narrowband signals (the reverse of the standard Doppler shifting argument)? Although the time scaling of the STV wavelet model replaces the frequency shifting operation of the previous approaches, the STV wavelet operator can still model frequency shifting. For *narrowband* signals, time scaling is nearly equivalent to frequency shifting; in the limit, as the narrowband signal becomes a sinusoid, time scaling and frequency shifting become equivalent. It is interesting to note that LTI systems cannot perform frequency shifting. If a complex exponential is input into a LTI system, then that same complex exponential comes out of the system (it may be multiplied by a complex constant but its frequency is not altered). Thus, systems that create new frequencies must be time-varying. The creation of new frequencies is natural for the time scaling operator. Next, an example of the STV wavelet operator's application is presented.

Examples of the STV Wavelet Operator

This section employs the STV wavelet operator to model a practical scattering problem. The physical model will be a rotating or spinning rod that has small reflecting spheres along its length. See Figure 5.7. This example can represent the reflection from the surface of an expanding or contracting heart. Contrary to the smooth heart surface considered previously, now the heart surface is considered as rough, so as to backscatter the incident signal all along the heart surface. For mapping velocities to scales, only the *radial* velocity of the heart surface toward the transmitter/receiver is considered. Since the heart surface is curved this radial (or projected) velocity in the direction of the waveform arrival will change for each location on the heart (the heart is assumed to expand or contract spherically for simplicity). At the "edge" of the heart (as viewed from the transmitter/receiver) the radial velocity will have no radial velocity component and thus the edge acts as the pivot point in this example (heart edge is at the pivot in Figure 5.7). As the point of incidence is moved away from the "edge" and toward the center of the heart, the radial velocity component increases; this corresponds to moving further out along the rod in the model in Figure 5.7.

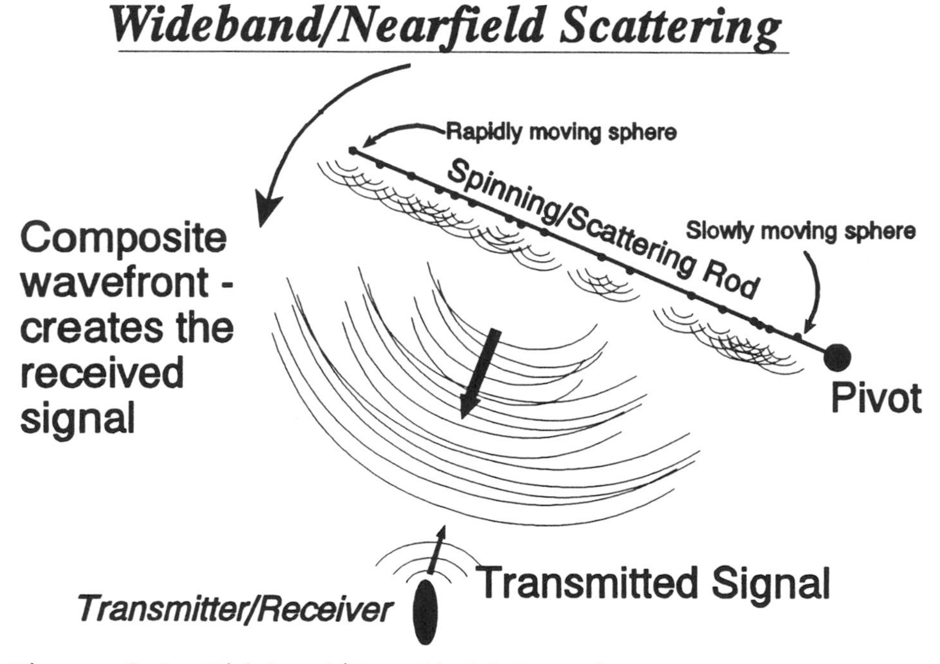

Figure 5.7: Wideband/Nearfield Example

The spinning rod and the environment through which the waves travel will be modelled as a system. The system input will be the transmitted waveform, while the system output will be the received signal (which is a composite sum of the many scattered *versions* of the transmitted signal). The transmitter and receiver are assumed to be collocated. Each reflected version of the transmitted signal is created just as the single reflection in the ultrasound heart example was created. Each separate scattering sphere along the length of the rod is treated as a separate "heart" in this analogy.

First, assume that the rod is motionless (not spinning). Assume also that each sphere has an exponential impulse response (the sphere is not rigid). Assume that the composite impulse response of the motionless rod (the reflected signal if a transmitted signal is an impulse function) is modelled as a decaying exponential function, $h(t) = e^{-t}$, $t > 0$. Obviously the models chosen are not exactly representative of the physical system that is shown; the intent is to provide a physical example for utilizing the STV wavelet operator and not create an exact model of the heart. Since this system is LTI (no motion) then the system's STV wideband system characterization, $P(a,b)$, will be the impulse response, a decaying exponential, on the unity scale line or $P(a,b) = \delta(a-1)\,h(b)$ as in Figure 5.8.

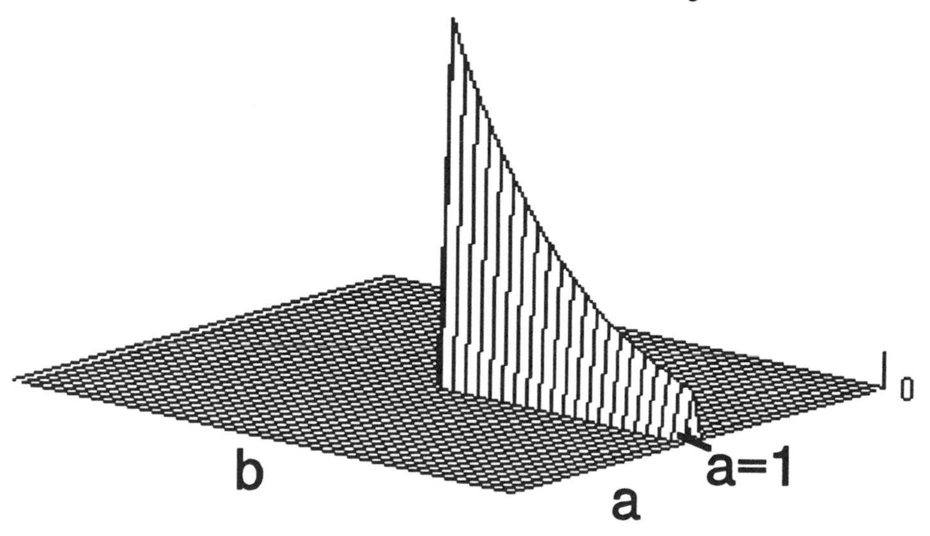

Figure 5.8: Wideband System Characterization of Time-Invariant System

Since this motionless system is time invariant, the STV wavelet operator collapses to just convolution and the output of the system (the received signal) is simply a convolution of the input (transmitted) signal with the system impulse response function (the LTI system model). Ignoring any delays in the system, the impulse response is modelled as a simple, real exponential, $h(t) = e^{-t}u(t)$, where $u(t)$ is the unit step function which is 1 for $t \geq 0$ and zero otherwise. If the transmitted signal or system input is also assumed to be an exponential, $x(t) = e^{-t}u(t)$, then the output, by convolution, is:

$$y(t) = recvd(t) = \boldsymbol{STV}_{\delta(a-1)\,h(b)}\,(x(t))$$
$$= \int_{-\infty}^{\infty} h(b)\; x(t-b)\; db = \int_{0}^{t} e^{-b} e^{b-t}\, db = t\,e^{-t} \qquad (5.24)$$

For comparison purposes all of the subsequent systems in this example also use a real decaying exponential signal, $x(t) = e^{-t}u(t)$ as the input signal.

The result of a simulation of the STV wavelet operator with the wideband system characterization shown in Figure 5.8 is the output signal presented in Figure 5.9.

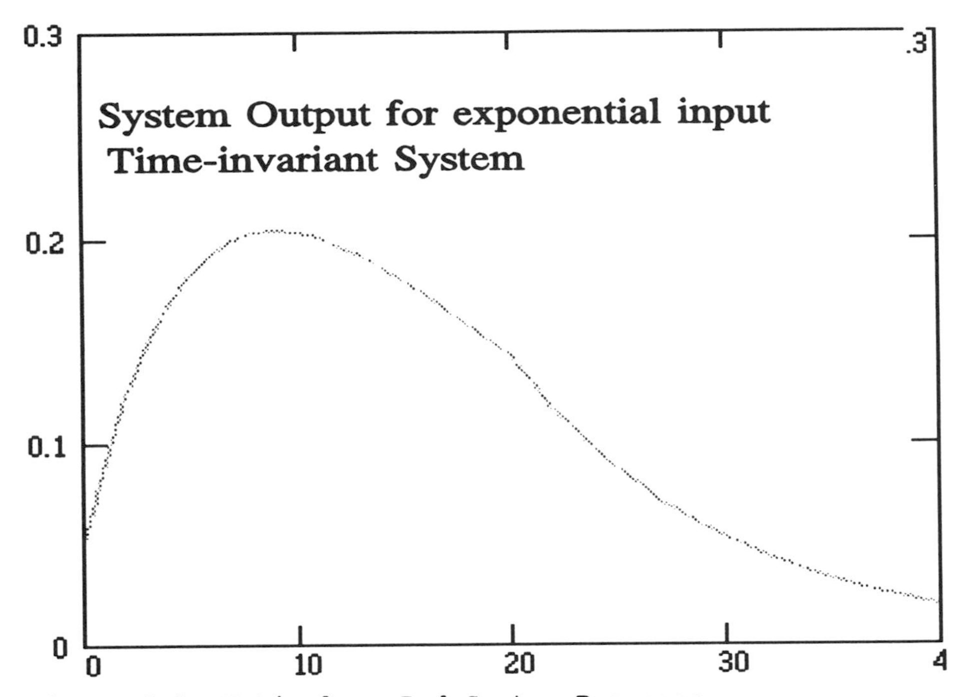

Figure 5.9: Motionless Rod System Response

This agrees with the convolution that would be performed by a LTI system model.

Next, assume that the rod of Figure 5.7 is moving and is rotating at a high rotational velocity in the counter-clockwise direction (toward the transmitter/receiver). Following the discussion of the ultrasound heart example, each scattered signal (a different scattered signal from each reflecting sphere along the rod) will be a compressed version of the transmitted signal. Due to the rotation, the linear speed of each scatterer (sphere) in the direction of the transmitter/receiver (radial speed) will increase as the position of the sphere increases away from the pivot point. The rod acts as a linear velocity multiplier. Thus, highly compressed versions of the transmitted signal will be reflected from the spheres near the far end of the rod (away from the pivot point), while only slightly compressed versions will be reflected from spheres near the pivot point. Observe that the time compression is not exactly a linear time compression; as the rod rotates the linear speed changes slightly and leads to acceleration terms. The acceleration terms are considered negligible in this example - a subsequent discussion considers the limitations imposed by acceleration(s). A LTI system with one particular impulse response cannot model the time scaling in this system. However, to a first order approximation of the time compression, the reflected signal can be modelled as a sum of scaled versions of the transmitted signal. The more general STV wavelet operator provides this model.

For this counter-clockwise spinning rod, a mathematically convenient approximation of the STV's wideband system characterization is a two-dimensional set of decaying exponentials, one at each scale and translation corresponding to the location of each scatterer. Mathematically, this wideband system characterization is defined as:

$$P(a,b) = \begin{cases} e^{-(b+0.1a)} & \text{for } b \geq 0 \text{ and } a \leq 1 \\ 0 & \text{elsewhere} \end{cases} \qquad (5.25)$$

This wideband system characterization is displayed in Figure 5.10.

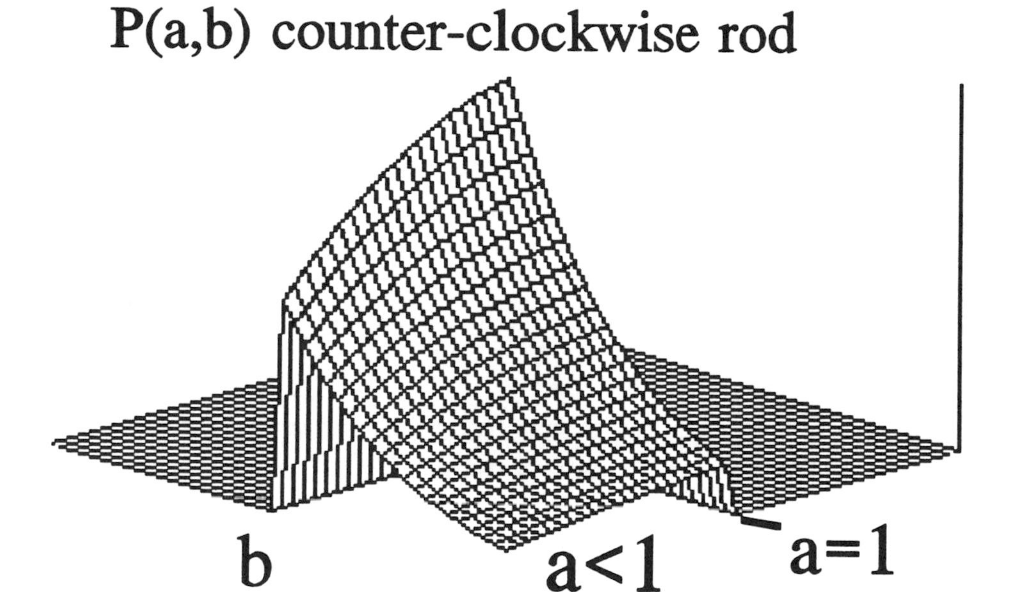

Figure 5.10: Wideband System Characterization: Counter-clockwise Rod

Again, this characterization is not an exact physical model; it is simply a wideband system characterization that has properties that are expected to be similar to those in this physical system. However, the physical model allows intuition to be applied to determine the expected qualities of the output (received) signal. Due to the scatterers all moving toward the transmitter/receiver, the system model should exclusively represent compressions and this is indicated by the wideband system characterization only having support for scale values less than one. Since this system is expect to only create compressed versions of the transmitted signal, the received signal is expected to have sharper changes and more variation than the received signal when the rod was motionless.

For the same exponential input signal as in motionless rod case, the system output can be obtained from the STV wavelet operator. The result of a simulation of the STV wavelet operator with the wideband system characterization of the rotating rod (equation (5.25)) is presented in Figure 5.11.

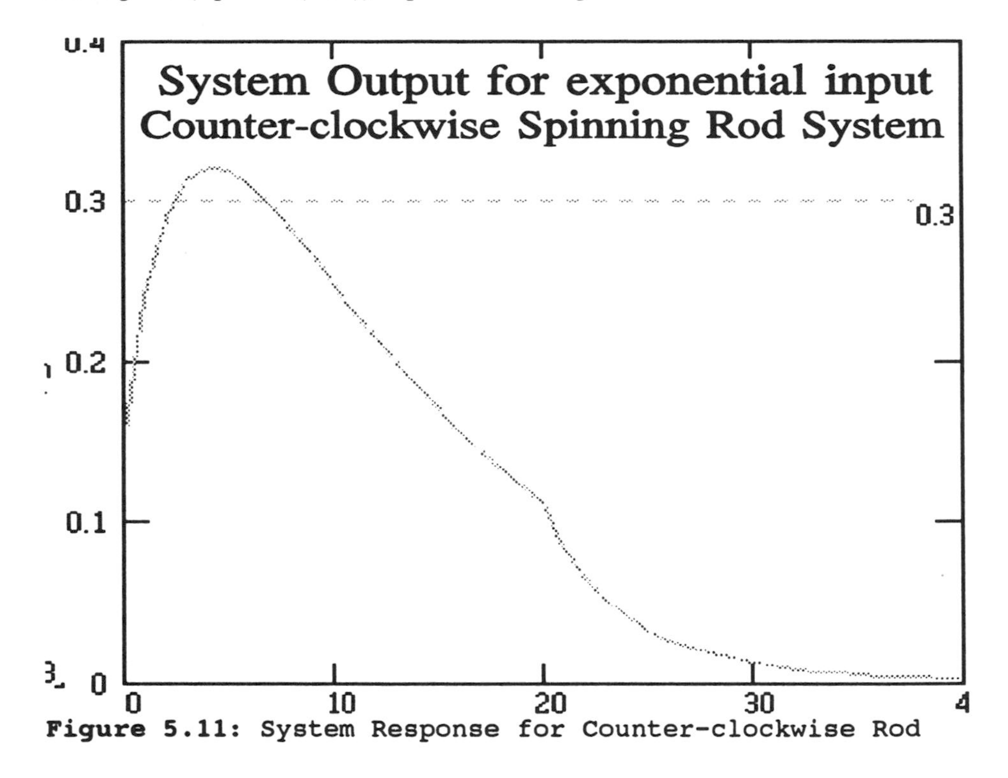

Figure 5.11: System Response for Counter-clockwise Rod

Repeating the aforementioned conditions but with the rod spinning in the clockwise direction (away from the transmitter/receiver), the received signal is a sum of versions of the transmitted signals that are expanded or dilated. For this situation an approximate wideband system characterization is provided in Figure 5.13. Note that only scale values larger than one have support, indicated a dilation of the input signal.

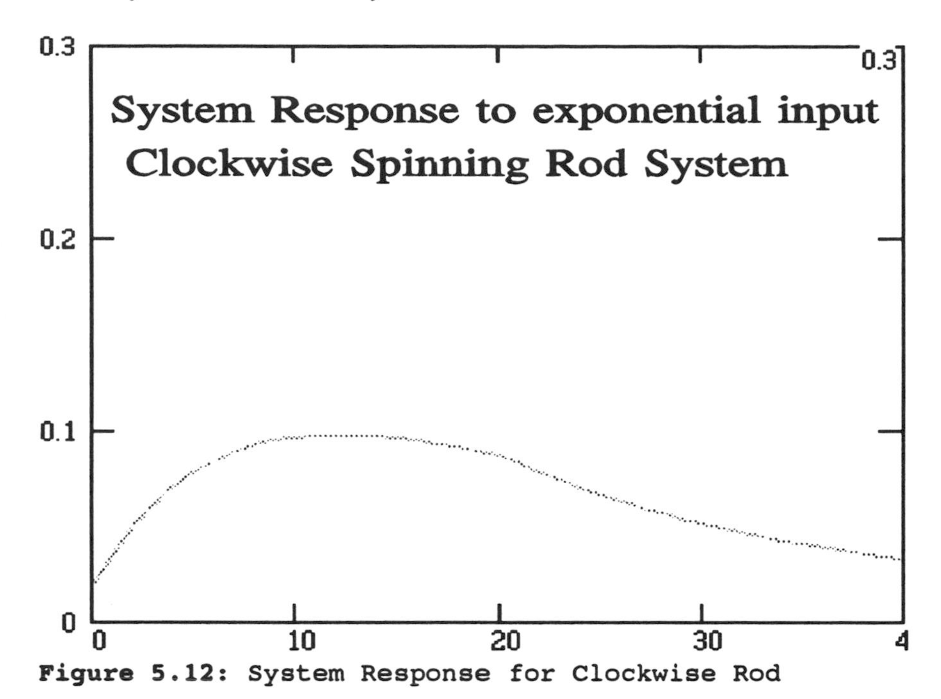

Figure 5.12: System Response for Clockwise Rod

With the same exponential input as in the other cases, the physical system model suggests that the received signal will be elongated and smoother due to the dilation of the reflected signals. The corresponding STV wavelet operator output signal demonstrates these properties and is shown in Figure 5.12.

As can be seen in Figure 5.9, Figure 5.11, and Figure 5.12, the output signals change as the wideband system characterization changes. Now consider a reformulated problem. Assume that the transmitted signal is known and the received signal can be measured. Can an accurate system model be deduced or estimated from these measurements? The qualities of the system can be partially deduced by comparing the input (transmitted signal) to the output (received signal). This is the system identification problem when it is done rigorously. These problems are briefly addressed in the rest of this chapter but Chapter 6 addresses the system identification more completely.

Wideband Reflection: Comparing the LTI and STV Models

At this point the STV wavelet operator and its properties and qualities have been presented. This section applies the new STV wavelet operator to the wideband reflection problem. Problems and limitations of LTI system models are identified and solved with the STV wavelet operator. Several of the STV wavelet operator properties are utilized.

Returning to the ultrasonic, heart reflection example discussed earlier in this chapter, it was assumed that a constant velocity, "head-on" reflection occurred. Thus, a scaled and delayed version of the transmitted signal is received when the reflector has non-negligible motion over the duration of the incident signal. A

Figure 5.13: Wideband System Characterization for Clockwise Rod

system model of this reflection process should be created for further analysis and to construct an estimator of the heart's location and motion (an image). What type of system is it? How does it react to various types of inputs? The most widely applied system model is a linear time invariant (LTI) system model. LTI systems lead to the convolution operation between the input signal and the system's impulse response. Earlier in this chapter it was demonstrated that time scaling (wideband reflection) cannot be properly modelled by a LTI system model. Another, more intuitive demonstration also exists.

This demonstration first considers a particular input signal and a corresponding system response (output signal) and finds corresponding LTI and STV system models. Then another input signal is applied to demonstrate the difference between these two system models.

The primary properties of LTI systems are that they are **linear** (the response to a weighted sum of different input signals is the weighted sum of the system response to each individual input) and **time-invariant** (the response of the system to a shifted input is the shifted response to an unshifted input - see Figure 5.14). These properties are used to establish a system model from a given input and output.

Assume the response to some input signal is known. From these input and output signals determine both a LTI and STV system model that will create the specified output from the given input. For example consider the input and output signals from Figure 5.2 on page 147.

Time-invariance Property

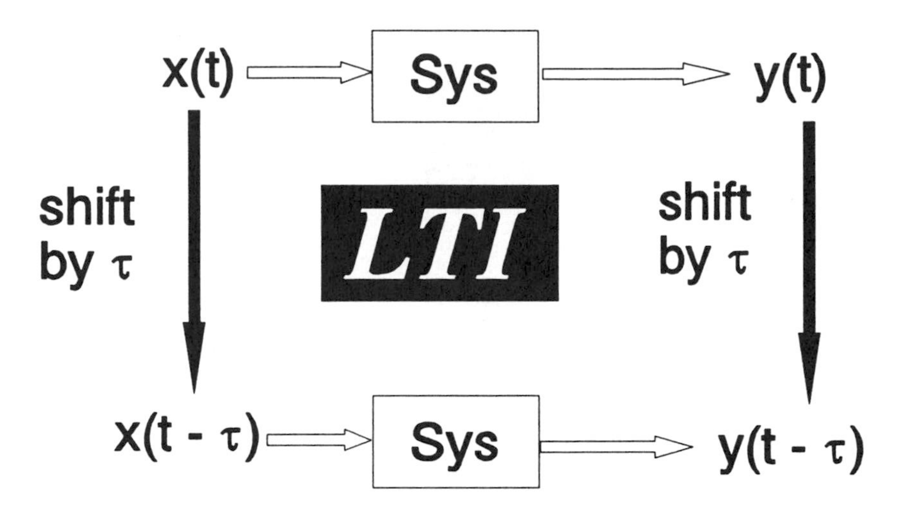

Figure 5.14: Time-invariance Property

For this input and output two system models can be established. If the system is assumed to be linear and time-invariant (LTI), then the output can be written as a weighted sum of two weighted and delayed versions of the input signal or $y(t) = 0.5\ x(t-3) + 0.5\ x(t-4)$. The corresponding impulse response is $h(t) = 0.5\,\delta(t-3) + 0.5\,\delta(t-4)$. If the system is instead modelled as being time-varying and performing time scaling, then the STV operator would have a wideband system characterization, $\mathbf{P}(a,b) = 2.82\ \delta(a-2)\ \delta(b-3)$. (Note that since the STV model includes the LTI model as a special case, it could model either one with different a set of assumptions). For this particular input and output signal pair, two possible system models exits; $h(t) = 0.5\,\delta(t-3) + 0.5\,\delta(t-4)$ and $\mathbf{P}(a,b) = 2.82\ \delta(a-2)\ \delta(b-3)$.

Now consider a new input as shown in Figure 5.15. For the LTI model and the impulse response determine previously, the response (output) to this new input can be written as a weighted sum of translated versions of this new input signal (the convolution of the impulse response and the new input). Figure 5.16 displays the output of the LTI system model for this new input. Note that the second portion of the new input is shaded, the response to the shaded portion of the new input is shown in Figure 5.16 as a shaded region; that response is added to the response of the other portion of the signal to create the LTI system output to the new input.

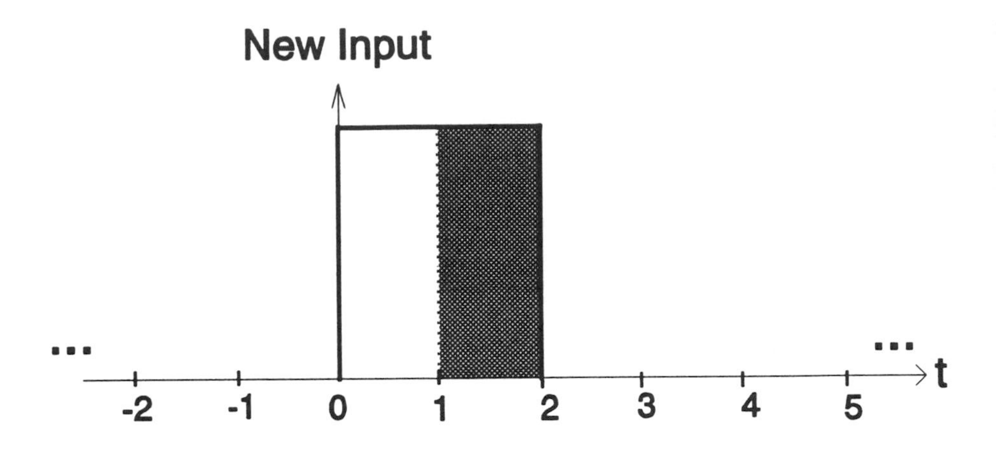

Figure 5.15: New Input Signal

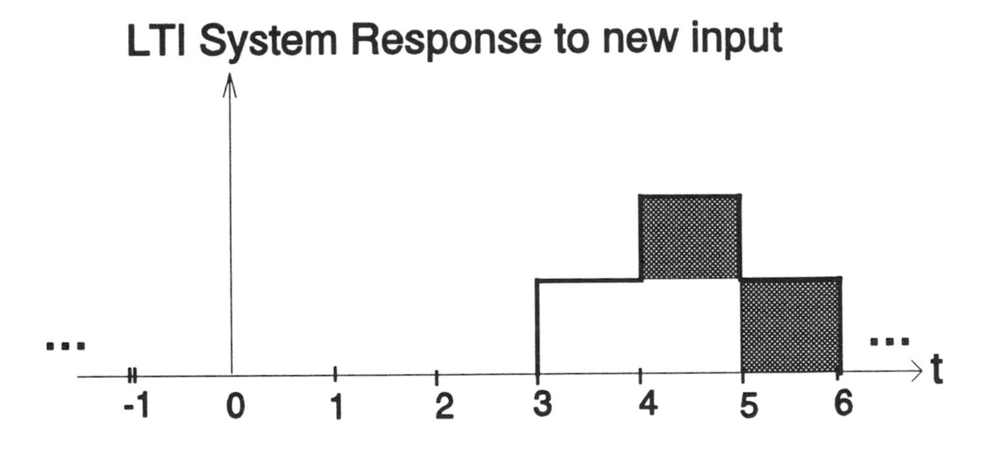

Figure 5.16: LTI Response to New Input

If the STV model with P(a,b) is applied to this same new input, then the new output signal will be a scaled version of this new input. For the wideband reflection model the reflection output for this new input is simply a scaled version of the original output signal as shown in Figure 5.17.

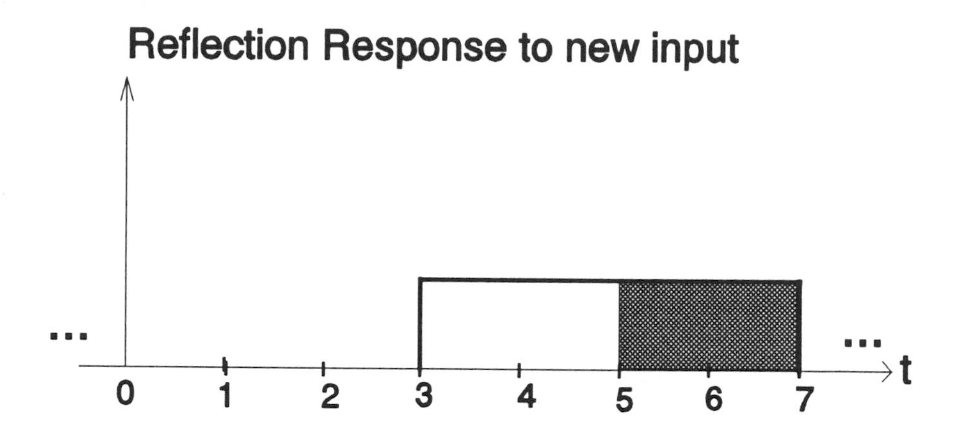

Figure 5.17: Reflection of New Input

Since the LTI model is a special case of the STV model, then two wideband system characterizations exist for this particular system: $P(a,b) = 2.82\,\delta(a-2)\,\delta(b-3)$ and $P(a,b) = \delta(a-1)\cdot[0.5\delta(b-3) + 0.5\delta(b-4)]$. Which one is correct? Both are obviously accurate models for the original input/output signal pair. Combining this previous example into one figure yields Figure 5.18. If additional input/output signal pairs were available, then the correct model could be resolved. However, an important distinction arises. When the system models are estimated the LTI estimation interval is typically "short" compared to the STV estimation interval. The LTI model assumes that the reflector has not "resolvably" moved over the processing or estimation interval (see Chapter 3) while the STV model accounts for the "resolvable" motion and allows the model to be valid over a longer processing interval. Assume that the input/output signals in Figure 5.18 are just the envelopes of a modulated signal (so that the signals have some bandwidth and an average value near zero). The LTI impulse response model would be estimated from the input/output signal pair over short observation intervals - if the output was observed for only 1 second intervals, then the output interval between 3 and 4 would look just like the input signal and so would the interval from 4 to 5 (if the time-varying impulse response models are used, then a frequency shift could also act on each of these 1 second packets). Thus, when processing "short" sections of the output signal the estimated system model is the two impulse LTI model. If, instead, the entire output interval from 3 to 5 is processed "as a whole," then the STV scaling operation matches a scaled version of the input signal to this output. So, for the wideband model a processing interval of 2 seconds is used whereas, in the LTI (or narrowband) model, the processing interval was only 1 second. As suggested in equation (5.1), if the processing interval is made short enough, then the LTI or narrowband model can always be employed. However, the shorter processing intervals can lead to poorer estimators. In conclusion, the LTI model is estimated if short estimation intervals are employed (even with the STV model) while the STV scaling model is estimated if longer estimation intervals are employed. In addition, the LTI system model would fail to properly represent the wideband reflection.

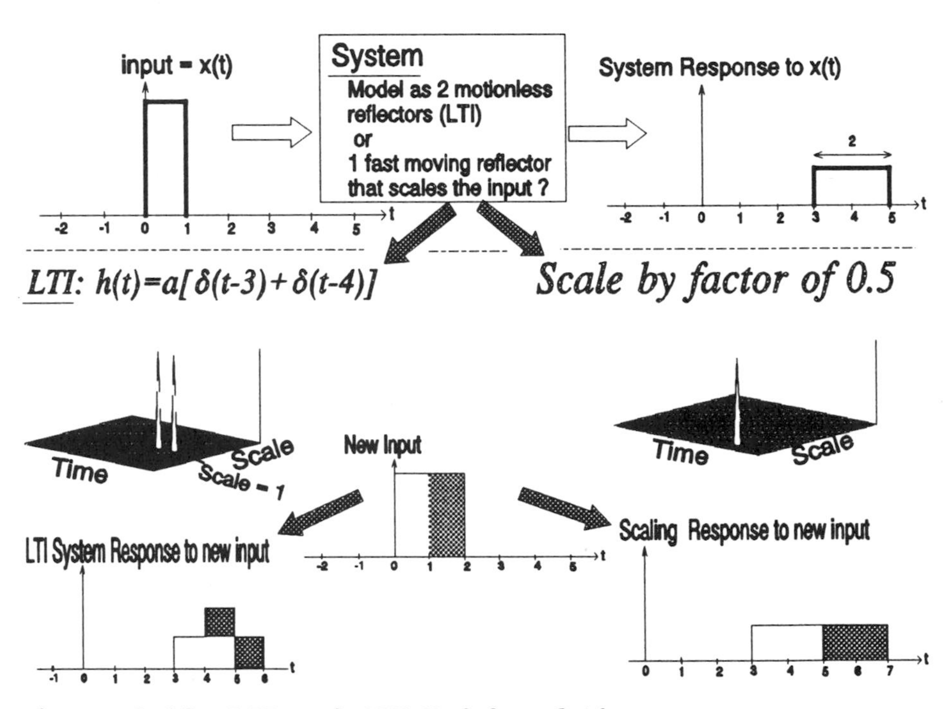

Figure 5.18: LTI and STV Models of the same Input/output Signal Pair

Justification for the STV Operator Instead of Convolution for Signals Represented by Wavelet Transforms

Reconsider the example in the previous section. Note that the new input is simply a scaled version of the original input. *The response of a LTI system to a scaled input signal is not a scaled output*; this is contrary to some, uncited research. *This "scaled input not causing a scaled output" condition of LTI systems is poor for modelling the system response to input signals modelled by wavelet transforms.* A wavelet transform decomposes a signal into scaled versions of the mother wavelet. If the response to each different scaled mother wavelet is different, then the linearity property cannot really be employed to advantage - the response to each different scaled mother wavelet must be separately characterized. A different system response for each scale value leads to a *scale-varying system* - a system that acts differently at each different scale value. For a input wideband signal, the interrelationships between different scale values (or bands, frequency octaves, etc.) at the input may be very different at the output (the interrelationship between wavelet coefficients of the output signal may be very different than the interrelationship between wavelet coefficients of the input signal). The changing response for each band can be interpreted as a *scale-aliasing*; this scale aliasing is blatantly displayed in the previous example where, in the LTI model, the response to a scale input signal caused overlaid output responses as in Figure 5.16. Since the scaling process necessarily shifts an exponential to a new frequency (and additionally changes the bandwidth of the signal) the scaling operation and LTI system models should not often be used together.

For signals characterized by wavelet transforms, the LTI system models (and really any of the convolutional approaches, as was previously demonstrated) are inadequate and, in general, should not be employed. A different system model should be employed to process signals represented as wavelet transforms. This system model is the STV wavelet operator; this section further motivates its existence.

Besides LTI system models, other linear system models have been proposed [Bos2, Pre, Zad, Zio] that are more appropriate for time-varying systems. However, estimating these system models for signals represented with wavelet transforms may be difficult as was demonstrated earlier in this chapter. For all of the aforementioned reasons the STV wavelet operator model should be considered as a model of systems that have significant motion involved in them and any input that is wideband and/or nonstationary.

The STV Wavelet Operator in the Wavelet Transform Domain

This section rigorously reformulates the STV wavelet operator in the wavelet transform domain. The input signal will be represented by a wavelet transform with respect to some arbitrary mother wavelet and the wavelet transform of the output signal (with respect to this same mother wavelet) will be the output of the system. Thus, this "wavelet domain system model" represents the action of the system in the wavelet domain *with respect to a particular mother wavelet.* The theory of several previous sections is employed in this section. Wideband ambiguity functions and their qualities are utilized to visually picture the multidimensional operations. The derivations implicitly employ the wavelet transform mother mapper operator but do not explicitly use it.

The STV wavelet operator maps a time (or space) domain signal to another time (or space) domain signal. If, instead, a "wavelet domain" STV wavelet operator could be formulated, then the input signal can be the wavelet domain coefficients and the new output would be the wavelet domain coefficients. An analogous composition of operators (frequency domain multiplication) was presented for LTI systems and the Fourier transform in equation (5.6).

Since the input is still $x(t)$ (but represented in the wavelet domain), the output should remain exactly the same; the wavelet transform domain representation should not affect the action that the system performs on the input signal. The output must not be affected if wavelet transforms are employed to represent the signals.

Since the system is being modelled by a STV wavelet operator, its definition will be the starting point of the derivation. But all that is required is the wavelet transform of the output *with respect to some arbitrary admissible mother wavelet,* $g(t)$. Transforming the output with respect to $g(t)$ yields and using the STV wavelet operator for $y(t)$ (and then rewriting $y(t)$ with the STV wavelet operator yields:

$$\begin{aligned} \mathbf{W}_g\,[y]\,(a_1,b_1) &= \int_{-\infty}^{\infty} y(t)\,\frac{1}{\sqrt{a_1}}\,g^*\!\left(\frac{t-b_1}{a_1}\right)dt \\ &= \mathbf{W}_g\,[c_x \mathbf{W}_x^{-1}\mathbf{P}(a,b)\,]\,(a_1,b_1) \\ &= \int_{-\infty}^{\infty}\left[\int_{-\infty}^{\infty}\int_{-\infty}^{\infty}\mathbf{P}(a,b)\,\frac{1}{\sqrt{a}}\,x\!\left(\frac{t-b}{a}\right)\frac{db\,da}{a^2}\right]\frac{1}{\sqrt{a_1}}\,g^*\!\left(\frac{t-b_1}{a_1}\right)dt \end{aligned} \tag{5.26}$$

Since the desired result is a mapping from the wavelet transform of the input, $\mathbf{W}_g x(\cdot,\cdot)$, to the wavelet transform of the output, $\mathbf{W}_g y(\cdot,\cdot)$, then the wavelet transform of the input must be formed under the integration. Rewrite a wavelet transform of a scaled and translated input signal, $\frac{1}{\sqrt{a}}\,x\!\left(\frac{t-b}{a}\right)$, as operations on the arguments of the wavelet transform:

$$\mathbf{W}_g\!\left[\frac{1}{\sqrt{a}}\,x\!\left(\frac{t-b}{a}\right)\right](a',b') = \int_{-\infty}^{\infty}\frac{1}{\sqrt{a}}\,x\!\left(\frac{t-b}{a}\right)\frac{1}{\sqrt{a'}}\,g^*\!\left(\frac{t-b'}{a'}\right)dt \tag{5.27}$$

then substituting a new time variable, $t' = \frac{t-b}{a}$, yields:

$$\begin{aligned} &\mathbf{W}_g\!\left[\frac{1}{\sqrt{a}}\,x\!\left(\frac{t-b}{a}\right)\right](a',b') = \\ &= \int_{-\infty}^{\infty}\frac{1}{\sqrt{a}}\,x(t')\,\frac{1}{\sqrt{a'}}\,g^*\!\left(\frac{at'+b-b'}{a'}\right)a\,dt' = \mathbf{W}_g x\!\left(\frac{a'}{a},\,\frac{b'-b}{a}\right) \end{aligned} \tag{5.28}$$

thus, writing the inverse of this transform as $\frac{1}{\sqrt{a}}\,x\!\left(\frac{t-b}{a}\right)$ and substituting back into equation (5.26) yields:

$$\mathbf{W}_g\,[y]\,(a_1,b_1) = \int_{-\infty}^{\infty}\left[\int_{-\infty}^{\infty}\int_{-\infty}^{\infty} \mathbf{P}(a,b)\left[\frac{1}{C_g}\int_{-\infty}^{\infty}\int_{-\infty}^{\infty} \mathbf{W}_g x\left(\frac{a'}{a},\frac{b'-b}{a}\right)\frac{1}{\sqrt{a'}}\,g\left(\frac{t-b'}{a'}\right)\frac{db'\,da'}{(a')^2}\right]\frac{db\,da}{a^2}\right]\frac{1}{\sqrt{a_1}}\,g^*\left(\frac{t-b_1}{a_1}\right)dt \tag{5.29}$$

now, changing the order of integration so that the time integration is the inner most operation, and only the functions that depend upon t are included under that integral:

$$\mathbf{W}_g\,[y]\,(a_1,b_1) = \int_{-\infty}^{\infty}\int_{-\infty}^{\infty} \mathbf{P}(a,b)\left[\frac{1}{C_g}\int_{-\infty}^{\infty}\int_{-\infty}^{\infty} \mathbf{W}_g x\left(\frac{a'}{a},\frac{b'-b}{a}\right)\left[\frac{1}{\sqrt{a_1 a'}}\int_{-\infty}^{\infty} g\left(\frac{t-b'}{a'}\right)g^*\left(\frac{t-b_1}{a_1}\right)dt\right]\frac{db'\,da'}{(a')^2}\right]\frac{db\,da}{a^2}$$

The integral over time is also a wavelet transform of a scaled and shifted mother wavelet, which has already been shown to be a wavelet transform that is shifted in both scale and translation, so:

$$\mathbf{W}_g\,[y]\,(a_1,b_1) = \int_{-\infty}^{\infty}\int_{-\infty}^{\infty} \mathbf{P}(a,b)\left[\frac{1}{C_g}\int_{-\infty}^{\infty}\int_{-\infty}^{\infty} \mathbf{W}_g x\left(\frac{a'}{a},\frac{b'-b}{a}\right)\cdot \mathbf{W}_g g\left(\frac{a_1}{a'},\frac{b_1-b'}{a'}\right)\frac{db'\,da'}{(a')^2}\right]\frac{db\,da}{a^2} \tag{5.30}$$

This is the **reformulation of the STV wavelet operator in the wavelet transform domain.** The input signal, $x(t)$, is wavelet transformed with respect to the mother wavelet, $g(t)$, and this wavelet transform, $\mathbf{W}_g x\,(a,b)$, is operated on by the wideband system characterization. This result requires some interpretation. First, recall that this entire equation is the wavelet transform of the output with respect to the mother wavelet, $g(t)$. Thus, the output can instead be created simply by applying an inverse wavelet transform to the left hand side of equation (5.30). But, remember that the STV wavelet operator is independent of any mother wavelet - the mother wavelet should not have any effect on the output. When equation (5.30) is examined, it looks as though the mother wavelet does affect the output. To resolve this difficulty, interpret the wideband auto-ambiguity function of the mother wavelet as the equivalent representation of an impulse function in the wavelet transform domain (mathematically, a reproducing kernel, see Appendix 2-A). Recalling the discussion and example in Appendix 2-A demonstrating the nonuniqueness of the wavelet transform domain representation of the mother wavelet, the wavelet transform of a mother wavelet with respect to that same mother wavelet is equivalent to an impulse function in the transform domain. Therefore, the wavelet transform of the mother wavelet with respect to the same mother wavelet acts as a sifting function in the transform domain. A problem arises with this "impulse" interpretation; impulses are not formed in the physical problems. When computed or estimated, the wideband auto-ambiguity functions of the mother wavelet have support across many values in the translation-scale plane instead of being just an impulse. The details of this computation are discussed in the next chapter when wideband scattering theory is detailed.

Ignoring the impulse interpretation of equation (5.30), another interpretation is with the narrowband scattering and ambiguity function interaction; this explanation is also detailed in the next chapter. But to provide some interpretation, consider the bracketed portion of equation (5.30) as a "smearing" effect analogous to the effects of windowing in spectrum analysis. In spectrum analysis the smearing or "ambiguity" is introduced in the transform or frequency domain as a convolution (smearing operator) with the spectrum of the window function (smearing function).

For the wideband analysis or wavelet domain, the mother wavelet is analogous to the window used in narrowband spectrum analysis. Instead of the smearing function being the spectrum of the window, it is now the wideband auto-ambiguity function of the mother wavelet (or the wavelet transform of the mother wavelet with respect to the mother wavelet, $\mathbf{W}_g g(\cdot,\cdot)$). The input transform, $\mathbf{W}_g x(\cdot,\cdot)$, is "smeared" by the wideband auto-ambiguity function of the mother wavelet before the system operates on this input (the wideband system characterization maps the input signal to the wavelet transform of the output).

Due to the smearing or coupling between the separate scales, a, a_1, a', *the effect of a wideband system characterization cannot be computed at each scale separately (its effect is nonseparable across the scale parameter). The output signal's wavelet coefficient at a particular scale depends on the input signal's wavelet coefficients at many scales.* The output's wavelet domain representation can*not* be constructed so that the only system effect is to multiply each wavelet domain coefficient by some constant that is a function of the mother wavelet and the original wideband system characterization. This invalid system operation (multiplication in the wavelet domain) would be analogous to the narrowband LTI system characterization in the frequency domain in which each of the output's spectral component at a particular frequency depends only on the input's spectral content at that frequency (the system's only effect is to weight these coefficients). This analogous wavelet domain operation is not valid for wideband/time-varying systems. Instead, wavelet domain smearing acts to *couple* the wavelet domain coefficients. When the mother wavelet is unconstrained, then a single wavelet coefficient of the output can depend upon many of the wavelet domain coefficients of the input.

For orthogonal wavelet transforms will such a structure be valid? In the orthogonal wavelet case the wideband auto ambiguity function, $\mathbf{W}_g g(\cdot,\cdot)$, becomes an impulse function; does the aforementioned "smearing" effect still exist? Will $\mathbf{W}_g g(\cdot,\cdot)$ just be an impulse function and allow the system to be represented by just wavelet domain coefficients that multiply the input's wavelet transform as in the narrowband case? Although the wavelets are orthogonal, they are only orthogonal on a preselected grid; thus, for the system operation to be represented by wavelet domain multiplies, the wideband system characterization, $\mathbf{P}(a,b)$, must also only exist on this preselected grid. Since the wideband system characterization is an unknown, this condition may be too constraining (impossible in general). Selecting the grid before estimating the wideband system characterization would require that all of the resolution requirements be a priori fixed (known) and that the mother wavelets be fixed as well. In many applications the wideband system characterization may need to be estimated on a very fine resolution grid. In applications where flexibility is required the unconstrained (nonorthogonal) wavelets may meet the desired resolution goals and maintain the flexibility to change as required (possibly adapt to changes in the systems being characterized). So even in situations where an "orthogonal wavelet" set is chosen, it may be nonorthogonal on the desired scale-translation domain grid.

Consider a particular orthogonal wavelet on a grid that is finer than the "orthogonal grid." Figure 5.19 displays an orthogonal wavelet constructed by I. Daubechies [Daub1]. When the scale step size is 2 and the translation step size is 1 (defining the scale-translation domain grid), the WBAAF of this orthogonal wavelet will be an impulse. When the scale-translation domain grid is made finer the WBAAF of this "orthogonal wavelet" is no longer an impulse. A picture of the WBAAF of a Daubechies' orthogonal wavelet on a finer grid is displayed in Figure 5.20.

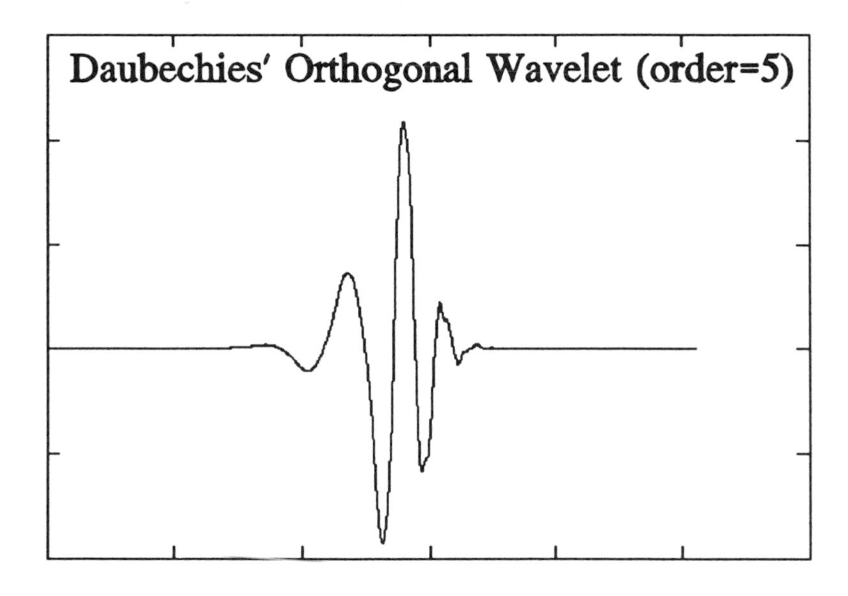

Figure 5.19: One of Daubechies' Orthogonal Wavelets

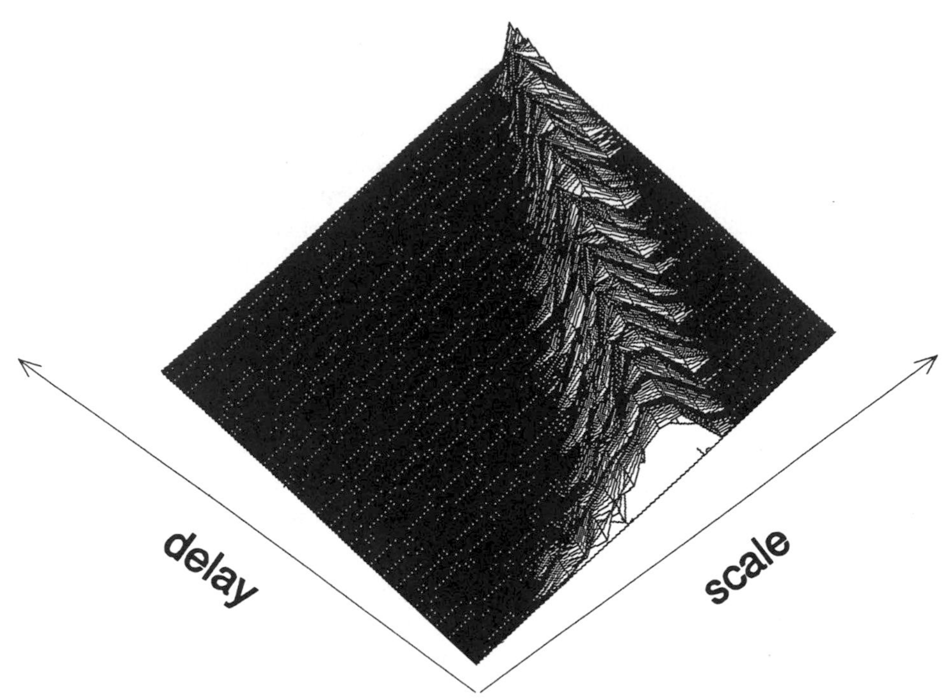

Figure 5.20: WBAAF of a Daubechies' Wavelet

As can be seen, the points of the WBAAF surface that are not exactly at the grid points are not orthogonal. The "orthogonal grid points" are still zero on this surface (except for the peak at the origin) but many other grid points have support. If the resolution is required to be finer than the "orthogonal grid," then the orthogonality condition is invalid on the new grid. The requirements for a finer resolution grid are discussed in general in the next section.

Space-time-varying System Identification Problem with Wideband/Nonstationary Input/Outputs

Identification or estimation of system models is used in many applications. For example, the modelling of a channel in communication or tracking problems. If a good model of the channel is available, then subsequent processing can be improved. Since many channels are time and/or space varying, present narrowband and stationary models are limited. The extension to space-time-varying systems models should improve the estimator performance in these applications.

In addition to a more general system model, more general input and output signals can be appropriately processed. Wideband and/or nonstationary input and output signals can be processed; either in the wavelet domain (as discussed in the previous section) or the appropriate time or space domain. With the efficient representations possible by using multiple mother wavelets to represent the signal and the efficient WBCAF generation (see Chapter 4) the processing can be reasonably implemented.

Besides the generality of these models and possible efficient representations, the system model requires fine resolution. Obviously, by equation (5.30) the mother wavelet is the limit of the achievable resolution in the system identification. But further "deconvolutional" techniques may improve the resolution even further. Simply selecting the peak of the magnitude surface is a "deconvolutional" technique that replace high regions (possibly the locations of auto ambiguity functions) with a single point. Although this is an efficient technique that provides high resolution system models, it is also quite sensitive to noise. Other "deconvolutional" techniques can be applied since the form of the mother wavelet's wideband auto ambiguity function is known (deterministic). These "deconvolutional" techniques include classical narrowband approaches [JohB] and wideband deconvolutional approaches [subject of current research].

The primary point is that high resolution estimates can be achieved. The scale resolution can be extremely fine. The power of 2 scales will most likely be unacceptable for the majority of the system identification problems. The fine resolution techniques of the general unconstrained techniques discussed in Chapter 4 appear to be appropriate for such applications.

Limitations of the STV Wavelet Operator - Time Referencing and Nonlinear Time Variations

Consider general time-varying or space-varying systems (the mathematics are consistently defined only with the time variable to avoid confusion, but several of the examples involve a spatial variation that is mapped into a time variation). These systems change across time; their response evolves with time. As can be seen in the ultrasound heart imaging example in Figure 5.1 and Figure 5.5, *the system response will be different at the next instant in time (half a heart beat later the heart is expanding rather than contracting and now the system model must change).* If the heart is expanding then the signal will be compressed rather than dilated, requiring a *new system model in this time interval.*

A time-varying system model must be referenced to some time - it is a valid model for that one time instant. However, a large interval of time can be used to estimate the system model at a particular time. The STV wavelet operator extends this valid processing duration.

Consider a system consisting a single reflector moving with a constant velocity directly at the collocated transmitter and receiver (the transmitted signal will be the system input and the received signal will be the system output). For one instant in time, the system can be characterized by a position and a velocity (a single point, a particular delay and scale, in the wideband system characterization). Obviously, at any time later (or earlier) the delay would be different; thus, a different wideband system characterization is required. However, the initial goal is not to estimate the system at every instant of time, instead the goal is to use a large interval of time to get a good estimate of the system at one instant in time.

For this example, the time scaling reflection model can be valid for a relatively long period of time, and that whole interval can be used to estimate the range and velocity or delay and scale of the reflector at one instant in time (chosen to be relative to the front edge of the transmitted signal). But, at some later time, the reflector finally reaches the transmitter/receiver and the model becomes invalid (relative to the observer or receiver, an acceleration has occurred - the closing velocity changed to a opening velocity). The time scaling model had some "time of validity" or valid processing interval that must expire at some point. For a reflector that is heading at some other angle rather than directly at the transmitter/receiver, an acceleration occurs much sooner and the "time of validity" of the system model is even shorter. When a processing interval becomes too long, a point reflector may lead to an invalid model that suggests an extended reflector; the wideband system characterization for an accelerating reflector will most likely include multiple points or a ridge but it should be a point).

Modelling the system as a time scaling operation has limitations. The limitations are analogous to those of the narrowband processing that were stated in Chapter 3. However, instead of the maximum velocity limiting the observation interval, now the maximum acceleration will limit the observation interval. Acceleration represents the rate of change of the velocity and velocity can be mapped to time scaling; thus, the acceleration represents the rate of change of the scale parameter. If the scale parameter is changing rapidly, then the gain or resolution achievable at a specific scale degrades. Although the system can be represented by these changing scales, this representation becomes sensitive and non-robust due to the large scale uncertainties caused by short valid processing intervals (and leading to low gain and poor resolution). *The systems that are modelled by the time scaling operation are well modelled only over time intervals during which the acceleration is small.* So, just as in the narrowband case, the modelling is good over some limited duration; however, the duration should be longer for this new model because it depends upon the acceleration rather than the velocity. Analogous to the narrowband valid processing interval, which was limited by the velocity and bandwidth (the system and signal variabilities), a valid wideband processing interval "bound" can be approximated. The general form is not true for all signals; it is a representative bound that can be qualitatively compared to the narrowband valid processing interval bound for comparison. The wideband valid processing interval bound for an accelerating system is approximately:

$$T_{p\text{-}WB} < \sqrt{\frac{c}{2 \cdot acc \cdot BW}} \tag{5.31}$$

As can be seen from this bound, the processing interval now depends only on the acceleration (an infinite processing interval can be used if only constant velocity exists). Also note that the processing interval decreases only as the square root of the reciprocal of the bandwidth-acceleration product. Thus, the valid processing interval for the wideband processor can be significant even when accelerations exist in the system. As previously noted, each different signal provides a different scale resolution and will thus produce a different valid processing interval bound but these are not detailed in this book [Alt, Kel, Swi]. Next, a further discussion of time referencing in time-varying systems is addressed.

The response of the system depends upon some chosen time reference (absolute time rather than just a time difference). Why is a system termed time-varying? First consider a system that is time-invariant. A time-invariant system definition is presented as a picture in Figure 5.14. The system response to a time shifted version of the original input signal is a time shifted version of the output corresponding to the original input signal. For time-varying systems, the response to a shifted input signal may not resemble a shifted version of the response to the original input, as was previously stated for the single reflector and the ultrasound heart imaging example - depending upon when the signal begins reflecting from the heart and the duration of the signal, it can be compressed, dilated, or both. Thus, the system model is valid for some absolute time - the system model can be different at each time.

Consider the position of an accelerating particle as a function of time. Let x(t) be the position as a function of time. Let x_o be the position of the particle at some initial time reference, t_o. In addition, an initial velocity of the particle at t_o, denoted v_o, must be specified. If the acceleration is denoted acc, then the position x(t) can be expressed as $x(t) = x_o + v_o(t - t_o) + acc\,(t - t_o)^2$. Although the velocity of the particle is continuously changing due to the acceleration, the initial conditions characterize the system with only a single velocity. For small time deviations in time away from the initial time, the $(t - t_o)^2$ term will be small. As this term increases it becomes the dominant term in the expression. If the position of the particle could be approximated by only a position and a velocity at each time instant, t_i, then the velocity should change as a function of time (modelling the acceleration). For the STV system model only delay and scale (or position and velocity, respectively) are modelled and, thus, the model must change as a function of time. This position/velocity model can be good over small time intervals and its applicability depends upon the duration of that interval and the degree of acceptable distortion. Since the processing duration dictates the gain and resolution characteristics of the system model, it is often desirable to increase the duration over which the model is valid.

Another example of a time-varying system is a capacitor that has a heat dependent capacitance. If the capacitor is in a computer and the computer heats up after power is applied, then the response of circuits with this capacitor in them changes as a function of how long the computer was on (and on many other environmental conditions). For the system model of the circuits with this capacitor, the time reference becomes the computer "turn-on-time."

For both the computer circuits and the heart imaging, the system response is different at each time (given the same input), a different system model is required at each time instant. So an input signal must be decomposed into "instants of absolute time" and the response for each instant is added together to form a composite output signal (this is analogous to the time-varying frequency response considered in equation (5.9)).

A completely general system model would require the wideband system characterization to be a function of time as well, $\boldsymbol{P}(a,b,t)$. The desirable feature of the new system characterization is that it is valid for longer time durations or intervals. If the wideband system characterization is not a function of time (as in the previous analysis) then it must have some particular time reference associated with it and should probably be denoted as $\boldsymbol{P}_{t_{ref}}(a,b)$ instead of just $\boldsymbol{P}(a,b)$.

Obviously, this time referencing complicates the analysis of the time-varying systems and adds more symbols to an already crowded notation. Therefore, the limitations of the wavelet based system model were deferred until after the model was presented.

The motivation for the new wavelet based system model is to create a system representation that is valid for a longer period of time. Given that the system model still must change as a function of time, how long can a system model be acceptably accurate? The new model increases the duration over which the system model can be properly used. If the system model is accurate over longer durations, then the gain and resolution properties for estimating the system model can be improved. In addition, wideband and nonstationary input and output signals can be properly processed.

Although the STV wavelet operator is presented as a time-varying system model, a particular system model, a single wideband system characterization, $\mathcal{P}(a,b)$, is strictly valid at only a *snapshot* in time. This is exactly analogous to the time-varying impulse response characterization - a new impulse response at each particular time. The concept of a new frequency response at each time has an intuitive difficulty - frequency responses are not local in time; some time interval is required to define a spectrum. The wavelet transform system formulation has elements that are localized in time.

Obviously, since the system is time-varying, it is only characterized at a particular snapshot in time; however, the processing interval over which this model is "acceptably" valid is extended beyond the narrowband-stationary model's interval of validity. If the time variation has any nonlinear terms (e.g., an acceleration) then the aforementioned STV model is valid only over some limited duration - the STV model considers only linear time variations (i.e. affine operations on the time variable). By changing the characteristics of the mother wavelet, higher order variations can be modelled, but this technique is not considered here and is the subject of current research.

Bi-wavelet System Representation: Time variation of the Time-varying System Model

Since the wideband system characterization, $\mathcal{P}(a,b,t)$, is a function of time, its time variation can also be examined. This analysis is analogous to the bi-frequency distribution for narrowband system models [Zad, Zio]. It will be demonstrated that translations at unity scale of the **bi-wavelet representation** correspond to constant velocity system or channel variations while support at scales other than unity indicates acceleration(s) in the system variations. Thus, the multidimensional wavelet transform with respect to the time variable of $\mathcal{P}(a,b,t)$ can be formed for $\mathcal{P}(a,b,t)$ at two different times. This new multidimensional (MD) wavelet transform remains two-dimensional and is termed the bi-wavelet system representation. The kernel of this new MD transform is just the wideband system characterization at the initial time (the time reference) which is also two-dimensional (2D).

Since the techniques in the previous section estimated the wideband system characterization at specific times, only the wideband system characterization at specific time references will be used as the input to the bi-wavelet system representation. The bi-wavelet system representation operates on a 2D wideband system characterization at time reference t_2 or $\boldsymbol{P}(a,b,t_2)$, which will be abbreviated $\boldsymbol{P}_2(a,b)$. The wideband system characterization at the reference time is used as the MD mother wavelet, and it is abbreviated $\boldsymbol{P}_r(a,b)$. The bi-wavelet system representation analyzes the commonalities between $\boldsymbol{P}_2(a,b)$ and $\boldsymbol{P}_r(a,b)$. It also characterizes the differences between these two representations; the differences indicate the acceleration(s) in the system (and higher order "accelerations," if they exist). The **bi-wavelet system representation** is here defined as:

$$BiW(s,\tau) = \iint \boldsymbol{P}_2(a,b)\, \boldsymbol{P}_r^*\!\left(\frac{a}{s}, b + \frac{a\tau}{s}\right) \frac{da\, db}{a^2} \qquad (5.32)$$

This equation is very close to the mother mapper operator. The interpretation is also very similar.

First consider a system with no motion or the bi-wavelet system representation for no time change ($\boldsymbol{P}_2(a,b) = \boldsymbol{P}_r(a,b)$). For this case $BiW(s,\tau) \approx \delta(s-1)\,\delta(\tau)$ because the two wideband system characterizations are the same. Thus, the bi-wavelet distribution has support only on unity scale, indicating the system changes only with a constant velocity (zero in this trivial case). When a constant velocity change occurs in the system, then at a later time the points in the wideband system characterization will still be at the same scale (velocity) but will be translated in the translation parameter (delay or range) - see Figure 5.21. For this figure, the bi-wavelet distribution will only have support along unity scale again (constant velocity system variations).

When the bi-wavelet representation has support only along unity scale, the dimensionality of the representation collapses. This is exactly analogous to the narrowband correlation functions; if the process is nonstationary, then two time variables are required, but when the process is stationary, then the dimensionality of the 2D time plane collapses to a line (representing the differences in time). Here, the "stationarity" of the model is constant velocity. If only constant velocities occur, then the time scaling operation is valid over the entire interval (stationary).

The bi-wavelet distribution can be compared to the bi-frequency distribution from narrowband theory [Zio]. The second frequency variable indicates the rate of change of the system's frequency response. If the frequency response does not change over time, then the second frequency will have support only at zero (d.c.). Thus, exactly analogous this narrowband bi-frequency representation, the bi-wavelet distribution's region of support indicates the type of variation in the system.

When *accelerations* occur, the wideband system characterization changes in both *scale* and translation. The bi-wavelet representation will have significant support only off of the unity scale line. The acceleration could correspond to a constant velocity reflector in the nearfield such that its radial velocity is constantly changing. For special cases such as this, the bi-wavelet scale parameter can be directly related to the acceleration (angle to the reflector). For reflectors that are

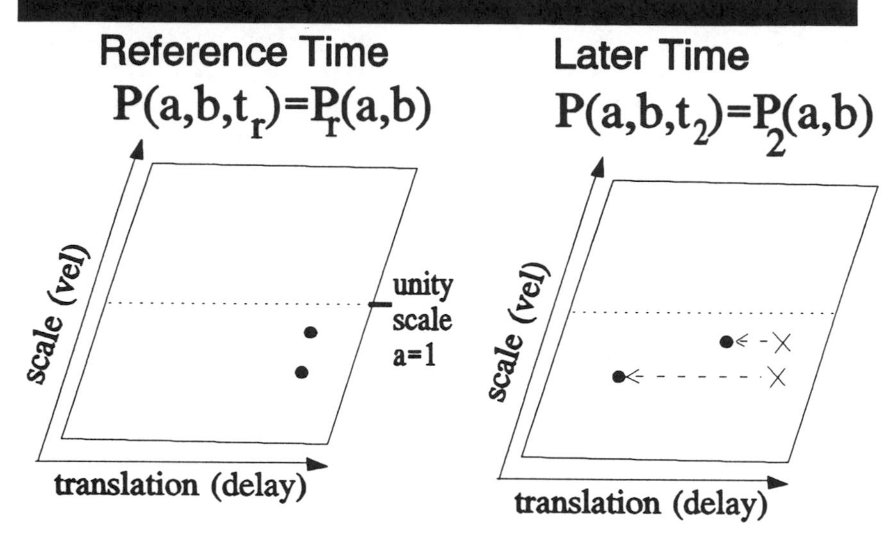

Figure 5.21: Inputs to Bi-Wavelet Processor

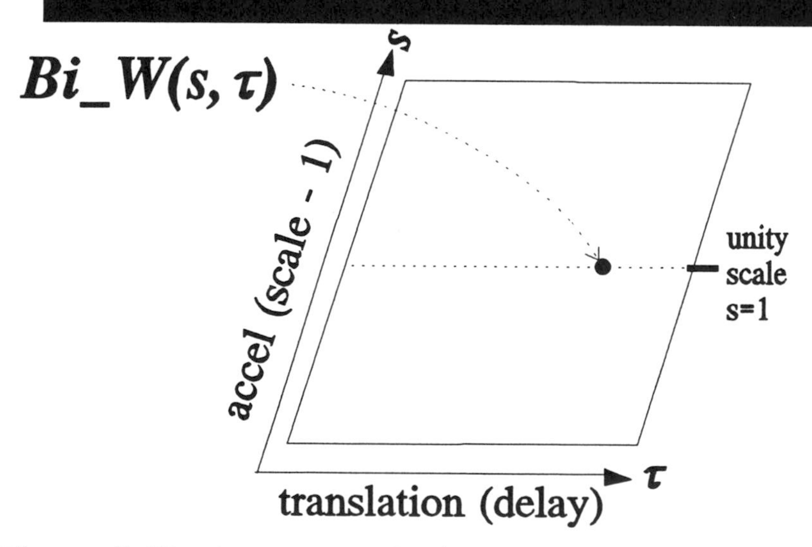

Figure 5.22: Constant Velocity System's Bi-Wavelet Distribution

both accelerating and traveling at an angle, the interpretation of the scale parameter in the bi-wavelet representation is difficult; however, any acceleration leads to support off of the unity scale line. Thus, the scale of the bi-wavelet distribution is related to system accelerations while the translation parameter of the bi-wavelet distribution is related to the constant velocity variations in the system.

Simple examples with one or two constant velocity reflectors that approach the transmitter both radially (directly toward) and non-radially (to the side of the transmitter) verify the concepts. More general wideband system characterizations can also be employed but the interpretation quickly becomes too complicated. The bi-wavelet distribution is still being refined and is the subject of current research.

The next chapter develops further applications of the STV wavelet operator and its wavelet domain reformulation to general wideband scattering theory. The wavelet domain formulation of the STV wavelet operator is the starting point of that analysis.

Chapter 6: Wideband Scattering and Environmental Imaging

Introduction

All of the mathematics and wavelet theory in this book can be applied to the problem of characterizing or imaging an environment. Environments can range from a single cell to the entire human body, the layers of earth to outer space, or the particle specks in a drop of water to the whole ocean. The characterization of these environments include modelling the position, motion, and other properties of objects or boundaries in the environment. In addition to identifying specific objects or features, a complete model of a channel or environment (including spreading, dispersion and any other environmental effects) may be desired. Assumptions regarding the environments can be enforced by parametric models (i.e., assume that only one narrowband source is in the environment and it is in the farfield of all sensors - then the characterization of the environment can be performed by closed form models and estimators).

As stated in Chapter 5, the new wideband/wavelet model attempts to estimate the environment (image) at one snapshot instant of time but processes the signals over a time interval (the estimation interval - often the length of the signal being processed). *It is assumed in the wideband assumption that the environment may have a different image or status at the end of the estimation interval than it did at the beginning of the estimation interval - the estimator accounts for the linear time variation over the estimation interval.* The narrowband models assume that the environment has not changed over the estimation interval. The narrowband assumption can become invalid over a shorter period of time than the wideband model. The extra estimation time achieved by employing the wideband model can lead to more robust estimates, greater energy, higher gains, better resolution and other desirable characteristics.

Many practical applications of both wavelet and ambiguity theory involve modelling the reflection, refraction, and/or transmission process. The combination of these processes is sometimes referred to as the *scattering process*. The scattering process can be interpreted as the action of the environment on the given input signal. The environment modifies a given input signal and creates a new signal that is received at the sensor. See Figure 6.1. For these reasons the environment is modelled as a system that acts on the given input signal. The system model is the wideband system characterization, $P(a, b)$, (detailed in Chapter 5) and the estimate modelled as a system that acts on the given input signal. The system model is the wideband system characterization, $P(a,b)$, (detailed in Chapter 5) and the estimate of this wideband system characterization becomes the estimate of the scattering process. The estimation of the wideband system characterization is the system identification problem; thus, the wideband system identification problem and the scattering function estimation problem are identical in this book's formulation. For system identification, a controlled (or known) signal is input into the system and the output signal is processed with the input signal to form a system model. In the scattering case the input signal is the transmitted signal. After obtaining an estimate of the scattering process it can then be mapped to an image of the environment. See Figure 6.2.

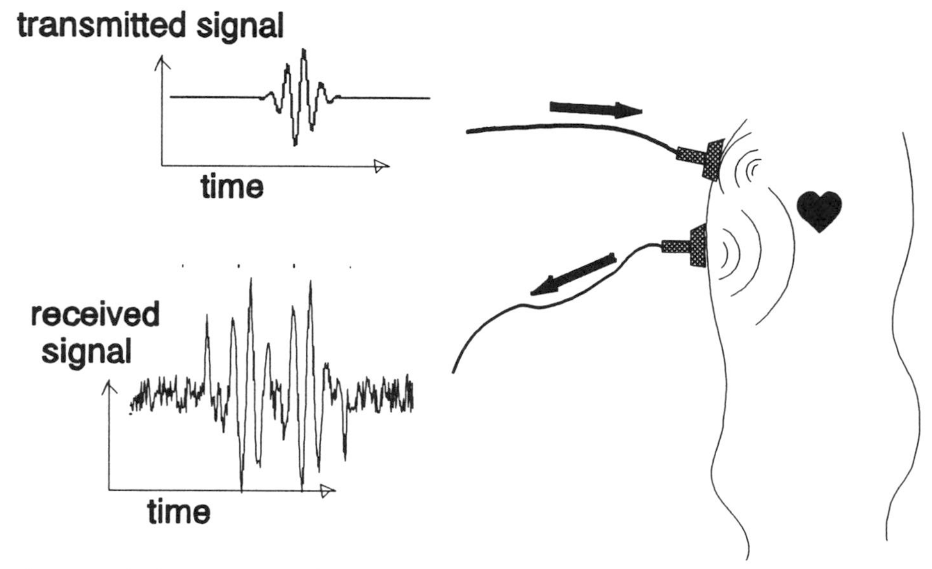

Figure 6.1: Scattering Function Estimation

This book concentrates on the signal processing aspects related to wavelet theory and avoids most of the physics involved with specific applications; this allows the presented research to be very general and widely applicable. To illustrate the application of wavelet theory and to demonstrate the constraints imposed by the physics of an application, a physical application of wavelet theory is addressed and is detailed later in this chapter.

One application is active sensing or imaging of an environment. Although scattering processes are most often interpreted with active sensing systems, the same wideband scattering concepts can be applied to passive sensing systems. The mathematics and algorithms are identical; only the interpretation and the physics changes. For passive sensing applications one of the received signals is simply treated as the transmitted signal in the active sensing system (one received signal is

Scattering Characterization to Image Estimation

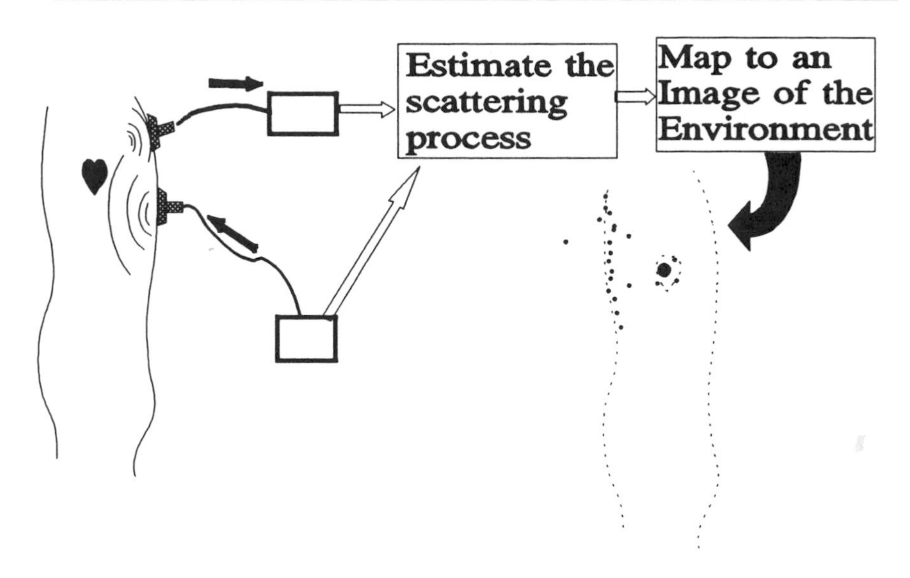

Figure 6.2: Imaging by Wideband Scattering

employed as the replica signal in the wideband matched filter processing). Thus, the mathematics will be general and applicable to either passive or active sensing systems.

For image processing, estimating or modelling the image blurring process is equivalent to estimating a system's scattering process. For this case the input signal is a known unblurred image and the scattering (or blurring) process acts to blur the image. With a blurred image and the unblurred input image, the blurring process can be estimated. One notable difference is that the new wideband scattering models are not constrained to be positive or real. The scattering model can be complex and represents the "action" of the system. This complex scattering function representation can be mapped to an energy distribution due to its interpretation as an inverse wavelet transform (see Chapter 5).

To avoid confusion, only one dimensional signals will be addressed. This is consistent with the analysis and algorithms presented in Chapter 5. The scattering function is the selected terminology to represent the environment even though the wideband system characterization is complex. An example of a scattering process is the signal transformation created when a person's body undergoes an ultrasound imaging exam (the heart imaging example of Figure 5.1 only included the heart and not the surrounding organs, fluids, etc. that make up the entire scattering process). The scattering function creates an output signal containing both desired echo signals (i.e. from the heart) and other, interfering signals from the transmitted input signal. The scattering process can be represented by several alternative representations. In this book the scattering process is represented as a *system* that acts on the input signal(s) to create the output signal(s). Thus, the linear system modelling in the previous chapter is extensively exploited in this chapter.

This chapter presents a wavelet theoretic model for the *wideband* scattering process. The ramifications, justifications, and limitations of this model are detailed. Initially, narrowband scattering theory is discussed to establish a starting point for the wideband processing and to display the limitations of the narrowband scattering theory. In the active sensing case, the narrowband scattering function can be mapped to estimate the range and motion of the scatterers [Van]. To determine the expected performance of these estimators, statistical assumptions and techniques are applied [Van]. This book does not thoroughly address the statistical performance analysis of wideband scattering; the scattering function estimators are stated but its performance is not detailed. Finally, wideband signal processing considerations for signal design are addressed

Scattering Theory

Critical to the design and analysis of radar, communication, medical imaging, and many other systems is the characterization of propagation channels and reflecting/emitting sources. Due to the uncertainty and randomness of these channels and sources, statistical models are required for meaningful characterization. The standard approach that has been used for characterization of stochastic channels has been the *scattering or uncertainty function* approach. Scattering functions model the second order statistics of stochastic channels and sources (the scattering process) and show how the energy is dispersed (ambiguous) in delay and scale or frequency. These parameters can be mapped to the position and movement of a scatterer. The scattering process can include point reflectors, distributed reflectors, or many other types of scatterers; an ultrasound scattering example was presented in Figure 5.1. For general problems the transmission process must also be assumed to have variations and inhomogeneities and these effects are sometimes included as part of the scattering process. The transmission process can also be modelled as a system and the parameters of that model can be estimated by system identification techniques discussed in the last chapter.

The classical, narrowband scattering function approach has several important limitations. First, it is based on the narrowband ambiguity function concepts that are often invalid if any of the following conditions are true: the signals have a large time-bandwidth product (and, practically, the signals have a moderate fractional bandwidth, about 10% or greater); the signals are low pass; or the signals result from nearfield scattering. Secondly, the most widely used scattering function approach uses the so called "wide-sense stationary, uncorrelated scattering (WSSUS)" assumption over some processing interval. It is important to relax these assumptions to achieve longer processing intervals that will lead to higher gains and better resolution. It can be shown that the wide-sense stationarity assumption implies that scattering is independent of absolute frequency (i.e., it is only a function of frequency differences). This assumption allows the *narrowband ambiguity function to be invariant with respect to frequency translations and is invalid for wideband signals and dispersive channels* (as will be subsequently demonstrated). In the active sensing case, a frequency difference characterizes the velocity of a scatterer. For the general reflection process the *frequency difference only approximates the signal's time scaling due to scatterer motion*. The frequency difference model is only valid for small scatterer velocities relative to the signal's velocity resolution [Ric1, Van]. This condition limits the processing interval over which this model is valid and, thus, limits the gain and resolution achievable with this model as well. When the motion exceeds these small limits then wideband processing is required. The wideband models will be valid over longer intervals and should provide higher gain and better resolution.

General Scattering Function Background

Standard narrowband processing uses the scattering function to estimate an environment [Bel, Van, Zio]. A signal is transmitted and then a new signal is received. The received signal is processed by the correlating it with time delayed and frequency shifted versions of the transmitted signal. This narrowband correlation processing is the narrowband matched filter presented in equation (3.14) and pictorially in Figure 3.16. The output of the matched filter is a surface in delay (range) and Doppler frequency shift (delay-Doppler plane). The narrowband scattering function can be estimated by simply taking the magnitude squared of this surface (range-Doppler map). If the received signal has a high signal-to-noise-ratio (SNR), then a refined estimate of the scattering function can be formed by deconvolution [JohB].

The scattering function estimate can then be used to estimate the range and radial velocity of a scatterer (or equivalently, the time delay and Doppler shift of the received signal) [Alt1, Kel, Van2]. The simple equations mapping the peaks of the delay-Doppler plane to the range and velocity of point scatterers are:

$$range = c \cdot delay, \quad vel \approx \frac{c \cdot \omega_D}{2\,\omega_c} \tag{6.1}$$

where c is the speed of sound, ω_c is the center frequency, and ω_D is the Doppler shift. The statistics (including variances) of range and velocity can be determined from the statistics of the delay-Doppler estimates. If the delay-Doppler estimates are simply the peaks of the magnitude of the output narrowband correlation surface (matched filter output), then an estimate of these variances can be obtained analytically [Van]. The estimates can be shown to be optimal if the noise is Gaussian and several other assumptions are satisfied. For the wideband case the mapping from translation and scale to range and radial velocity was discussed in Chapter 1.

The matched filter's output surface formed from the received signal and narrowband replicas of the transmitted signal can be modelled as the scattering function being convolved with the input signal's auto ambiguity function [Bel, Van, Zio] (this relationship is discussed later in this chapter). Similarly, it will be demonstrated that an exactly analogous situation holds for the wideband scattering. For narrowband scattering, this result establishes the requirements and structure that lead to optimal signal design [Van, Zio]. For the narrowband case, the NBCAF is invariant with respect to translations in both variables and, thus, the signal design problem is formulated about the origin without loss of any generality. The form of the narrowband ambiguity function and the convolution operation representing the scattering process are well accepted results and are discussed in the next section to provide an analogy for developing the wideband theory.

Narrowband Scattering Theory

Before formulating the wideband scattering theory, the narrowband scattering theory is briefly presented. The analogies between the narrowband and wideband scattering are exploited. Narrowband cross correlation or matched filter processing is addressed. Under several assumptions, the estimation of the range and velocity of point scatterers in the environment is performed by estimating the delay-Doppler peaks in the correlator output. These peaks are the estimate of the narrowband scattering function. For the active sensing case, the signals to be correlated are the transmitted signal and the received signal (see Figure 6.2). The received signal will be modelled as the output of a time-varying linear system when the input is the transmitted signal.

This discussion of narrowband scattering theory relates the modelling to the previously presented linear system models. Since it is assumed that the scatterers can be moving, then this environment (the position of the scatterers) is changing over time, or is time varying. A time varying system model is required and the system model is chosen to be the time-varying system model presented in equation (5.11). The system or channel is modelled with the superposition integral and the *time-varying impulse response*, $h(t,\tau)$. This general linear filter maps the input (transmitted) signal to the output (received) signal. The time-varying impulse response is the response of the system to an impulse at time τ; the response to an impulse changes across time, t.

Now, if received signal is correlated with the transmitted signal, then the second order statistics of this correlator output can be examined. Since the statistical properties of the transmitted signal are assumed to be known, only the statistics of the "system" or environment affect the correlator output statistics. Thus, the statistics of the system (really the time-varying impulse response, $h(t,\tau)$) must be examined. The correlation or second order statistic of this system model is:

$$\mathbf{E}\{h(t_1,\tau')\,h^*(t_2,\tau'')\} = \mathbf{R}_h(t_2,t_1,\tau'',\tau') \tag{6.2}$$

Under the assumption that the signal returns from scatterers at different ranges are uncorrelated, then the correlation of the time-varying impulse response at two different delays (ranges) is:

$$\mathbf{R}_h(t_2,t_1,\tau'',\tau') = \mathbf{R}_h(t_1,t_2,\Delta\tau)\,\delta(\Delta\tau)\,;\ \text{where}\ \Delta\tau=\tau''-\tau' \tag{6.3}$$

A different assumption can be imposed on the system or channel model. If the system or channel is assumed to be *wide-sense stationary or time-invariant* in the time parameter, t, *over the transmitted signal duration*, then the correlation depends only on the time difference [Bel, Pap, Ric, Van, Zio]. Instead of representing the system with the time-varying impulse response, the system is represented in the frequency domain by the time-varying frequency response (equation (5.10)). Note that the time-varying frequency response is usually referred to as the **spreading function** in the active sensing literature [Van, Zio]:

$$\mathcal{S}(\omega,\tau) = \int h(t,\tau)\, e^{-j\omega t}\, dt$$

The spreading function represents the *rate of change* of the impulse response with its frequency variable. If the system did not change across time (time-invariant), then the spreading function would only have support at zero frequency ($\omega = 0$) and would be zero everywhere else. When the time-varying impulse response is WSS, then the spreading function will be dependent only on the time differences; the distribution over frequency will be the same for the same time differences. So the correlation of the spreading function for the WSS system is:

$$\begin{aligned} E[S(\omega_1,\tau')\,S^*(\omega_2,\tau'')] &= E[S(\omega_1,\tau')\,S^*(\omega_1+\Delta\omega,\tau'')] \\ &= E[S(0,\tau')\,S^*(\Delta\omega,\tau'')]\,\delta(\Delta\omega) \quad \text{where } \Delta\omega = \omega_2-\omega_1 \end{aligned} \tag{6.4}$$

The system is still time-varying but the time variation is only a function of time difference; thus, the correlation can have support at any frequency, but its support is the same for the same frequency differences (indicated by the impulse function). Now the correlation of the spreading function is dependent only upon frequency differences and not absolute frequencies.

This WSS assumption is made in many narrowband processing systems. Even systems that claim to track time variations or nonstationarities of the environment often assume a "local" or short stationarity period, so that the WSSUS system model is valid over that "small" interval. Obviously, if this interval is required to be small, then the processing gains, resolution and other performance measures can be correspondingly small.

When the system, channel, and/or environment is assumed to be wide-sense stationary (WSS) in time and has uncorrelated scattering (US) in delay, then the "system" is denoted as a **wide-sense stationary, uncorrelated scattering (WSSUS)** system. The **narrowband scattering function,** $\mathcal{NBS}(\Delta\omega,\Delta\tau)$, is defined as the correlation of the WSSUS spreading function. Due to the WSSUS assumptions the second order statistics of the spreading function (the narrowband scattering function) depend only on the difference in the two delay variables and the difference in the two frequency variables. This dependence only on differences says that the WSSUS assumptions cause the second order statistics to be the same everywhere in the delay-frequency plane; *the WSSUS assumptions cause the second order statistics to be invariant to both time and frequency shifts*:

$$\begin{aligned} E[S(\omega+\Delta\omega,\tau+\Delta\tau)\,S^*(\omega,\tau)] &= \mathcal{NBS}(\omega,\tau)\,\delta(\omega-\Delta\omega)\,\delta(\tau-\Delta\tau) \\ \text{thus; } \mathcal{NBS}(\omega,\tau) &= E[|S(\omega,\tau)|^2] \end{aligned} \tag{6.5}$$

This narrowband scattering function involves an expectation for the WSSUS system's correlation function. The expectation is approximated by time averaging under an ergodic assumption [Pap1].

For the active sensing case, a narrowband correlation receiver creates a NBCAF between a received signal and a replica (transmitted) signal. Under the WSSUS assumption, the magnitude squared of the receiver's output surface can be represented by the convolution of the input signal's ambiguity function and the environment's *narrowband scattering function.* The magnitude squared of the correlation receiver's output surface function is:

$$|\mathbf{NB\_Corr}_{output}(\omega,\tau)|^2 =$$
$$= \iint \mathbf{NBS}(\omega',\tau')\; |\mathbf{NBAAF}_{input}(\omega-\omega',\tau-\tau')|^2 \frac{d\omega'}{2\pi}\, d\tau' \tag{6.6}$$

As an illustrative example, choose a narrowband signal with a narrowband auto-ambiguity function (NBAAF) shown in Figure 6.3. Assume that the environment is WSS in time and consists of three uncorrelated point scatterers. This environment can be characterized by three double Dirac delta functions (the WSSUS scattering function for this environment) in the two-dimensional time-frequency plane as shown in Figure 6.3. Then, following equation (6.6), the expected output of a narrowband correlation receiver would be that shown in Figure 6.3. Essentially, the ambiguity function is perfectly reproduced at the position of each of the scattering function's impulses. This convolution model is used extensively in simulations and performance analysis models. Signal design theory includes the examination and manipulation of the interactions between ambiguity functions themselves and also their interactions with scattering functions [Alt, Van].

Reconsider the problem of imaging the environment. Peaks of the correlation receiver output (in the delay-Doppler plane) can be mapped directly to range and velocity estimates of the scatterers (equation (6.1)). The resolution and statistical performance analysis of these estimates has been extensively researched [Van, Zio, and many others]. The uncertainty in the position estimate corresponds to the uncertainty in the estimation of the correlation receiver output peaks. Thus, for the uncorrelated, stationary narrowband case, the uncertainty in the position of a point scatterer is dependent only upon the transmitted signal's ambiguity function. Obviously, interfering sources are not considered here. Thus, *the uncertainty in the estimation of the correlation surface peaks is dependent only on the transmitted signal and is independent of the location of the scatterer in the time-frequency plane (independent of the velocity of the scatterer); this will not be true for the nonstationary wideband case.*

The WSSUS correlation receiver output formulation in equation (6.6) is the basis of the signal design problem. The signal design problem is concerned with optimizing the signal qualities to extract the desired information or image an environment. The signal can be designed to be mismatched [Van] with the interfering features in the environment. The interfering features of the environment

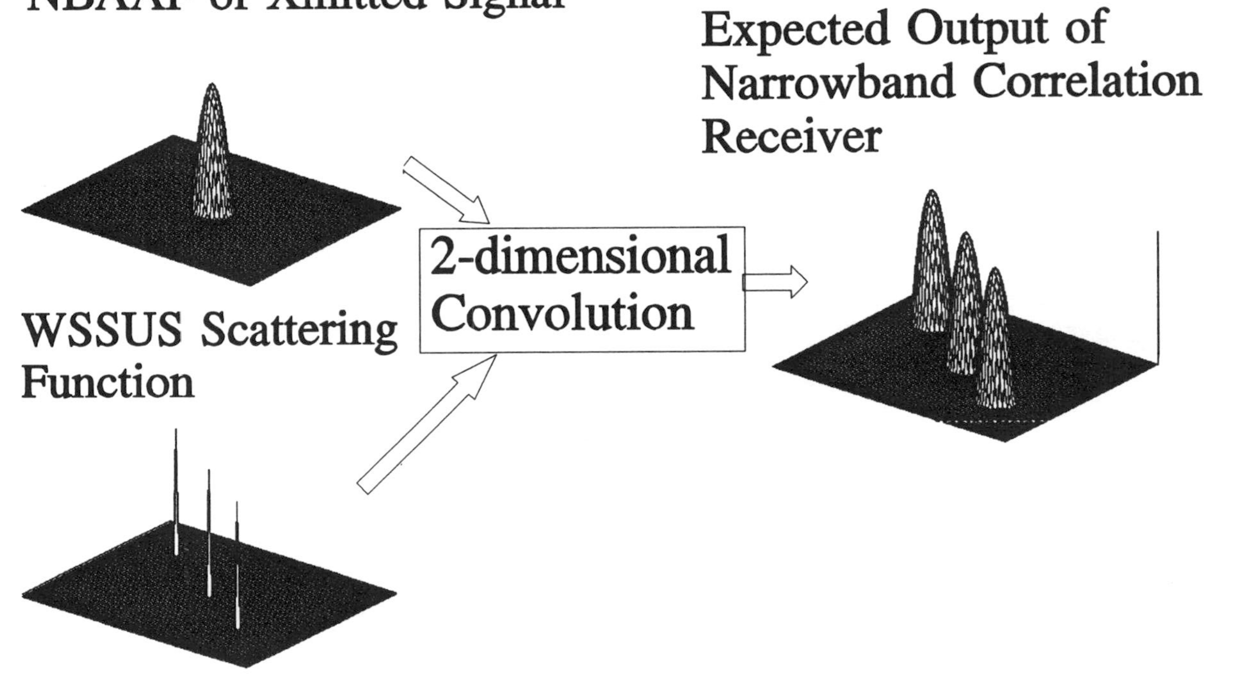

Figure 6.3: NB Correlation Receiver Output

can be characterized by an interference scattering function and then the signal is designed to have an ambiguity function that minimally overlaps with this interference scattering function [Alt1, Kni, Spa, Van2]. The characteristics of the correlation receiver output can be partially controlled by this design of the transmitted signal. If the scattering function is known, or can be partially estimated, then the signal can be designed to "properly" interact with this scattering function to achieve the "desired results" which are usually signal to interference gains [Alt, Van, Zio].

Ambiguity function analysis is used to determine the performance (variance) of a maximum likelihood parameter estimator [Alt, Van] that estimates the positions of the scatterers in the time-frequency plane. The estimator essentially picks the peak(s) of the magnitude of the correlation receiver output. The performance or variance analysis of the maximum likelihood estimator is determined by the moments of the ambiguity function around the origin, $\tau = 0$, $\omega = 0$ (for the wideband ambiguity function the origin is at $b = 0$, $a = 1$). Due to the translation *invariance* of the narrowband ambiguity function across the entire time-frequency plane (the ambiguity functions looks exactly the same at any point), variance analysis around the origin is the same as it is around any other point in the plane. That is not true for the wideband ambiguity function.

Wideband Correlation Receiver and its Output

The primary operation performed in this chapter is the cross correlation processing between the received signal and replicas of the transmitted signal. Referring back to the diagram of the matched filter in Figure 3.9 on page 93, this cross correlation processing can be termed matched filter processing. The output of the wideband matched filter processing is a surface in the delay and scale plane. An example of the magnitude of the matched filter output was shown in Figure 2.10. The connections between WBCAFs and wavelet transforms has been established in Chapter 3.

In relating the wideband scattering to the well established narrowband scattering, the same notation will be employed. Essentially, wideband ambiguity functions replace narrowband ambiguity functions, wavelet transforms replace Fourier transforms, and the wideband scattering functions replace the narrowband scattering functions.

Analogous to the narrowband correlation receiver a wideband correlation receiver is formed by hypothesizing many different time scales and delays and creating replicas of the transmitted signal at these delays and scales. This correlation receiver is essentially a wavelet transform; however, after reformulation it more closely resembles an inverse wavelet transform. By examining the correlation receiver's output and relating it to the STV wavelet operator, the analogies to the narrowband correlation receiver and narrowband scattering theory become clearer.

First consider the active case. Assume that the transmitted signal is x(t). The wideband correlator is:

$$Corr(a',b') \triangleq \sqrt{a'}\int_0^T y(t)\, x^*(a't-b')\, dt \tag{6.7}$$

Modelling the received signal, y(t), with the STV wavelet operator yields a reformulated wideband correlation receiver output:

$$\begin{aligned} Corr(a',b') &= \sqrt{a'}\int_0^T \mathbf{STV}_{WBS(a,b)}(x(t))\, x^*(a't-b')\, dt \\ &= \sqrt{a'}\, c_x \int_{-\infty}^{\infty}\int_{-\infty}^{\infty} \mathbf{WBS}(a,b)\left[\frac{1}{\sqrt{a}}\int_0^T x\left(\frac{t-b}{a}\right) x^*(a't-b')\, dt\right]\frac{db\,da}{a^2} \end{aligned} \tag{6.8}$$

or rewriting with the wideband auto-ambiguity function of the input signal:

$$\begin{aligned} Corr(a',b') &= \\ &= c_x \int_{-\infty}^{\infty}\int_{-\infty}^{\infty} \mathbf{WBS}(a,b)\, \mathbf{WBAAF}_{a,b}x\,(a',b')\, \frac{db\ da}{a^{5/2}} \end{aligned} \tag{6.9}$$

where the operators were previously defined and $\mathbf{WBS}(a,b)$ is the **wideband scattering function** (analogous to the wideband system characterization function, $\mathbf{P}(a,b)$, that describes the variations (scatterers) in the system). See Figure 6.4 for a pictorial demonstration of equation (6.9). *Note that in equation (6.9) the wideband auto ambiguity function has a subscript,* (a,b) *, that indicates the center of the wideband ambiguity function and dictates the shape of the particular ambiguity function as in Figure 6.4.* Actually, the shape of the ambiguity function is the same (invariant) for translations in delay but it changes as the scale, a', changes. *The wideband ambiguity function is not invariant with respect to scale changes - a different shape occurs for each different scale.* Thus, for scatterers with different scale values, the shape of the correlator receiver output will be different. (In Figure 6.4 the amplitudes in the wideband correlator output surface appear to be too low when compared to the auto-ambiguity function, but they are simply viewed from a different angle to emphasize the different shapes of the warped ambiguity functions; the un-normalized output surface is shown in Figure 6.5.)

The wideband correlator receiver output formulation of equation (6.9) simplifies the interpretation of wideband signal processing, performance for environmental imaging. The wideband correlation receiver's output characterization in equation (6.9) can be identically interpreted as the narrowband correlation receiver's output characterization in equation (6.6). Thus, the wideband signal design problem and the statistical performance analysis will concentrate on equation (6.9) just as narrowband signal design concentrated on equation (6.6).

For estimators of the range and velocity of a scatterer (or equivalently, the time delay and scale of the received signal) the second order statistics (variances) of range and velocity are determined. These range-velocity variances are determined from the statistics of the delay-scale estimates. If the delay-scale estimates are simply the peaks of the magnitude of the output wideband correlation surface, then

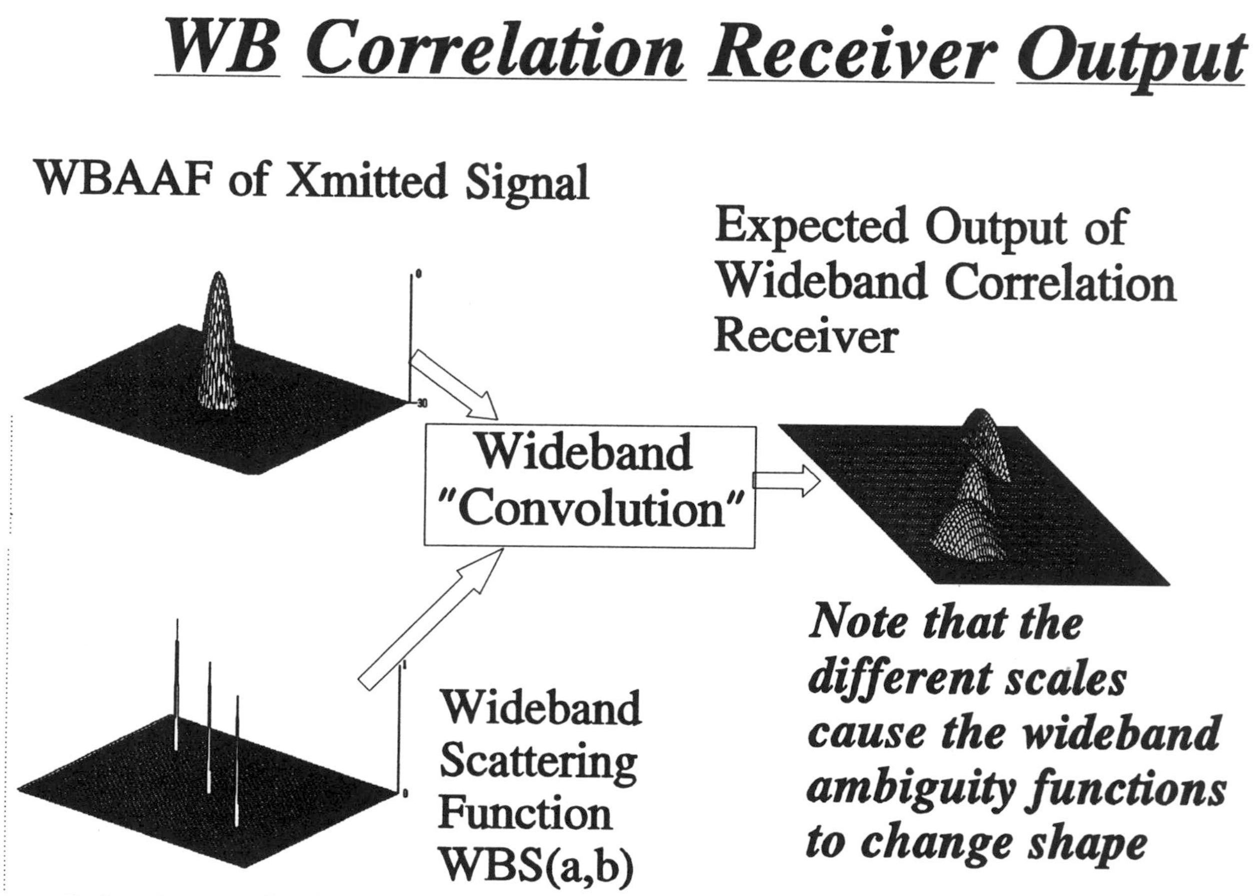

WB Correlation Receiver Output

WBAAF of Xmitted Signal

Expected Output of Wideband Correlation Receiver

Wideband "Convolution"

Wideband Scattering Function WBS(a,b)

Note that the different scales cause the wideband ambiguity functions to change shape

Figure 6.4: WB Correlation Receiver's Expected Output

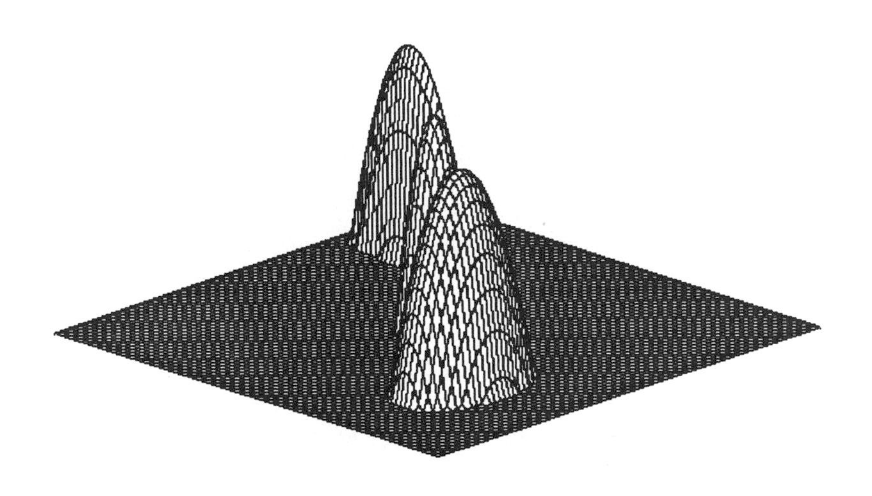

Figure 6.5: WB Correlator Output from Same View as other Surfaces

an estimate of these variances can be obtained analytically [Alt]. The wideband correlation surface between the received signal and wideband replicas of the transmitted signal can be modelled as the scattering function being "convolved" with a "*warped version*" of the input signal's ambiguity function (obviously the operation is not standard two-dimensional convolution due to the scaling operation, but the analogous interpretation exists).

Wideband Point Scatterer Example

Since the ambiguity function can be applied in many areas, the specific application of the ambiguity function must be made clear. For imaging an environment, such as the internals of a body, the ambiguity function has two specific applications. First, for both the wideband and narrowband cases, the cross correlation receiver (or matched filter) can be viewed as a cross ambiguity function between the received signal and the transmitted signal. Secondly, the auto ambiguity function of the transmitted signal determines the resolution and gain properties of the correlation receiver.

If the range and velocity of the dominant scatterers and/or emitters is desired, then the peaks of the magnitude of the wideband correlation receiver output should be determined. Under the point scatter and/or point emitter assumption(s), the peaks of the magnitude of this output surface are the desired estimates of the scattering function. Under several further simplifying assumptions (i.e., an isotropic, homogeneous medium), these peaks may map to position and velocity estimates (creating a physical image of the environment). The equations are stated later in this chapter after a specific correlator structure is chosen.

To simplify the problem being addressed in this section, the following assumptions are made. The medium is assumed to: be two-dimensional; have isotropic, spherical (circular) spreading; have a constant velocity profile; and a low Gaussian noise background ("low" compared to the signal energy). Observe that the assumptions do not include stationarity or time invariance of the scattering process. For the narrowband active case the details and equations for mapping from a NBCAF peak to a positional estimate were previously provided. For the wideband case the spatial mapping will depend upon the chosen form of the wideband ambiguity or correlation processor; this is subsequently detailed. The goal of this chapter is to interpret the characteristics of the processing and not to detail every step in mapping to spatial coordinates - these general characterizations work for other applications besides range-velocity estimation. Now that the motivation for finding a peak of the WBCAF has been established, consider the meaning and *estimation* of that peak.

Consider only a single peak in the wideband correlation receiver output. A single peak in the correlation receiver's output corresponds to a single point scatterer in the environment (i.e., a hard, highly reflective lump on the surface of the heart as in the ultrasound example). Since a point scatterer exists only at one point, it corresponds to a single range and velocity (at each instant in time the range becomes different, so a particular range is really only valid for an instant in time). Since the scatterer is assumed to be in motion, the reflected signal (and, thus, a portion of the received signal too) is a scaled and delayed version of the transmitted signal. In the image of the environment formed from the received signal, it would be desirable to see only an impulse at the correct range (delay) and velocity (scale). However, in forming the receiver output, an impulse is not generated. Instead, referring to equation (6.9), a wideband auto ambiguity function of the transmitted signal is created by the correlation receiver. This WBAAF will be at a point in the delay-scale plane corresponding to the range and velocity of the scatterer and will be appropriately warped according to the scale value.

For physical signals the wideband auto ambiguity function is not an impulse and, thus, has non-zero support at many scale and translation values. Obviously, these multiple scale and translation values map to multiple ranges and velocities. Since the true image was just a single point or single range-velocity, the estimated image is ***ambiguous*** ! Multiple ranges and velocities are indicated even though only one scatterer at one range and velocity value exists. The auto ambiguity function of the transmitted signal *at the correct scale* describes the ambiguity or uncertainty in the positional estimate. Note that the peak of this "warped" WBAAF is at the location corresponding to the correct impulse function - that is why the peaks on the wideband correlation receiver output are employed for position estimates.

A quick review of the last paragraph is necessary. First consider the *auto* ambiguity function of the transmitted signal. Its peak is always at a unity scale and zero translation ($a=1$, $b=0$). However, the wideband correlation receiver's output is described by scaled and translated ("warped") versions of this auto ambiguity function as in equation (6.9). These warped ambiguity functions have different shapes than the auto ambiguity function if the scales are not unity. If only unity scales are allowed, then the "warping" is only a translation (no scaling occurs) and the new ambiguity functions are just shifted versions of the WBAAF. This is exactly analogous to the narrowband case where the narrowband ambiguity functions anywhere in the delay-Doppler plane are just translated versions of the single NBAAF. This is intuitively acceptable because for scattering processes the Doppler frequency shifting is just an approximation of the time scaling. This approximation is valid only when the time scaling is approximately unity. When the scatterers start moving rapidly the narrowband and wideband cases will differ. Besides the analogies to the narrowband scattering, the wideband scattering resolution properties must be examined and are considered next.

Wideband Scattering Functions and Resolutions

Returning to the single point scatterer, the scattering function of a point reflector is an impulse in the delay-frequency (narrowband case) or delay-scale (wideband) domains. Choosing the peak of the magnitude squared of the correlation receiver output as the estimate of the scattering function can be sensitive when the signals are contaminated with interfering noise energy. When additive noise contaminates the received signal, the second order statistics of the peak estimation are still governed by the resolution properties of the transmitted signal. Figure 2.10 is a pictorial description this statement. Due to the WBCAF's form, the magnitude surface can be approximated as a quadratic surface near the peak. The noise will be superimposed on this surface and may cause the correlation receiver's peak to be at a location other than the peak of the "warped" auto ambiguity function. The ambiguity function describes the uncertainty in the peak estimate. If the ambiguity function rolls off quickly (has a sharp main lobe), then the correlation receiver's peak estimator should have a lower variance; obviously, any side lobes (extraneous peaks) would increase the variance. However, the sharper the peak, the more receiver correlations that must be formed; thus, trading off resolution and implementation efficiency as well.

Since the environment can have multiple scatterers, possibly very close together, the resolution ability of the correlation receiver is also important. Since the resolution properties are controlled by the wideband auto ambiguity function of the transmitted signal, wideband signal design must be considered as well as the interaction between the "warped" ambiguity functions and the wideband scattering function (e.g., to design optimal mismatch filters, etc. [Alt, Spa, Van, Zio]). Now the warping causes the wideband signal design problem to be different than the narrowband signal design problem. Improving the resolution of the correlation receiver's output beyond the WBAAF is a topic of future research but this has been done in the narrowband case. In the narrowband case the environmental estimate (image) can be improved beyond the narrowband ambiguity function resolution by *deconvolving* the correlation receiver's output surface with the input signal's NBAAF [JohB]. The environmental estimate is a refined estimate of the *scattering function*. Current research has formulated the analogous *wideband* deconvolution processor.

For wideband processing the resolution of the environmental scatterers depends not only on the input signal, but also on its interaction with the environment. Unlike narrowband scattering for which the narrowband ambiguity function of the input signal determines all of the resolution properties, the resolution properties of a wideband signal change due to their interactions with the environment. Instead of the ambiguity function simply being convolved with the scattering function as in the narrowband case, the wideband ambiguity function must change its shape depending upon the characteristics of the scatterers. Thus, *a single wideband signal resolves different portions of the environment with different resolutions.* Refer back to Figure 2.8 and Figure 3.3; the resolution is different for each different *scale* value. Reconsider the example in Chapter 5, once with the heart surface closing on the receiver and again with the heart surface going away. The resolution characteristics of the reflected signals change significantly. The shorter duration pulse in time, caused by the closing motion, will provide better time resolution, while the longer duration pulse, caused by the heart contraction, will provide better frequency resolution - even though the transmitted signal was identical in both cases. Refer back to Figure 6.4 to observe these different resolutions at different scale values.

Reconsider Figure 6.3 and Figure 6.4. Note the obvious difference; for the wideband case the resolution of the various scatterers in the environment changes as the position of these scatterers in the two-dimensional time-scale plane changes. Thus, *depending upon the characteristics of the environment, the resolution of the environment's scatterers changes.* In the narrowband case the resolution of the environment's scatterers is dependent *only* upon the characteristics of the signal and was *independent of the environment.* The independence condition is required for classical signal design.

As an aside, consider that the narrowband interpretation of the correlation receiver's output is that a copy of the ambiguity function is placed at the position of each scatterer in the time-frequency plane. If the wideband correlation receiver output in the right side of Figure 6.4 is interpreted with the standard narrowband interpretation, then the each "scaled" scatterer would appear as *distributed* scatterer. They must be interpreted as distributed scatterers because the resolution of the output correlation is finer than the transmitted signal's ambiguity function; finer resolution can only occur due to destructive interference (coherent interactions) of multiple ambiguity functions. If the scatterer was an extended scatterer, then multiple coherent ambiguity functions would interact and could possibly cause destructive interference to create finer resolution than the transmitted signal's ambiguity function.

Expectation and Stationarity

The definition of the narrowband scattering function required the expectation operation. The only way that the expectation operator is approximated in practical problems is by averaging over time. Since averaging approximates expectation only under the assumption that the signals are ergodic and stationary, the stationary condition (in delay and frequency - WSSUS) is required to form the narrowband scattering function estimates. Although only a "local" stationarity assumption is made, this is still the assumption over the coherent processing interval [Pap2, Pre]. If the stationarity assumption is invalid (or if the "local" stationarity interval is too short to achieve adequate gain, noise immunity, or resolution), then the "standard" narrowband expectation cannot be approximated with practical operations on the signals. For wideband signals and systems, as discussed in Chapter 3, the stationary conditions are invalid.

For wideband signals and systems the wideband scattering function is estimated by a cross correlation and any subsequent processing (i.e., wideband deconvolutional processing or other high resolution estimators). However, as discussed in Chapter 5, the system model or wideband system characterization (identical to the wideband scattering function) is estimated at a snapshot in time, over some interval of time. The "stationarity" assumption that is required is that the scale (velocity) is not changing over the estimation interval; the model only accounts for linear time variations. The estimate over time is not an approximation of the expectation operator as it is in the narrowband case; the estimate over time is performed to account for the time variation or scaling. The variance of these estimators will change with each different scale hypothesis and, thus, and expectation operator would yield different results for each different scale. A detailed statistical analysis of wideband correlators is not performed in this book but the variance of the estimators of delay and scale will be intuitively interpreted by analogy to the narrowband processing in the section after this next section. The next section places these estimators into a physical space.

Physical Form of the WBCAF

The wideband cross ambiguity function (WBCAF) was formulated as:

$$\mathbf{WBCAF}(a,b) = \sqrt{a}\int_0^T r(t)\, s^*(at-b)\, dt \qquad (6.10)$$

where r(t) can be the received signal and s(t) can be the transmitted signal. Besides the formulation of the WBCAF in equation (6.10), another WBCAF formulation (acts identically as a wavelet transform) is:

$$\mathbf{WBCAF}_{\text{not used}}(a,b) = \sqrt{a}\int_0^T r(t)\, s^*(a(t-b))\, dt \qquad (6.11)$$

This second formulation can obviously be mathematically mapped from the other formulation by using a substitution of variables. However, the wavelet transform (alternate WBCAF) in equation (6.11) will map directly to physical, spatial coordinates whereas the WBCAF in equation (6.10) would require an additional "warping" of its delay parameter by the estimated scale parameter before it could be placed into physical, spatial (range) coordinates.

As discussed in Chapter 3 several forms of the WBCAF exist. Only the WBCAF defined in equation (6.10) is used in this book. The primary reason for using this form of the WBCAF was because it was different than the wavelet transform (the WBCAF form in equation (6.11) acts just like a wavelet transform).

and to emphasize this difference. The difference between the wavelet transform and the WBCAF in equation (6.10) is the order in which the delay and scale operations are applied; for a wavelet transform the delay acts first and for the WBCAF the scale acts first. Also, for some applications in passive sensing, the implementation of this WBCAF may be more efficient due to the difficulties with creating scaled signals.

The difference between the wavelet transform and the WBCAF are more emphatically demonstrated by computing both of them for a specific signal and observing the characteristics of their outputs. The result is that the WBCAF unnaturally couples the delay and scale parameters while the wavelet transform processor keeps the delay and scale parameters uncoupled. Thus, for delay parameter to map directly to a range, the wavelet transform must be used rather than the WBCAF. This difference is detailed after considering a particular case in which the different processors all act nearly equivalently.

As noted previously, the order of the time scaling and delaying is the difference between the wavelet transform and the WBCAF. However, time scaling is always relative to some reference time; if the reference time is always chosen so that the delay is "small," then the difference between the wavelet transform and the WBCAF can be negligible. When passive sensing is performed and a data stream is broken into "blocks," then each block is a particular delayed signal (thus, delay acts first). However, each data block is then processed by a wavelet transform or a WBCAF; if the delays in these blocks are "small" compared to the block or bulk delay, then the two processor will provide similar results because the scale only acts on this small portion of the overall signal delay. Thus, if the WBCAF provides efficiencies in implementation, then it may be a better choice.

As justification of the differences, consider generating both a wavelet transform (or other WBCAF in equation (6.11)) and a WBCAF of the same signal. Consider a received signal (presented in Figure 6.6) that is just three scaled and delayed versions of a single function (employed as the second signal or mother wavelet). The wavelet transform of this composite signal is presented in Figure 6.7. Note that each ridge in scale and delay is a straight line parallel to one of the axis lines; only one scale or delay value is clearly indicated (no coupling exists between the delay and range parameters). Now consider the WBCAF of the same signal with the mother wavelet being the second signal in Figure 6.6. This WBCAF surface is shown in Figure 6.8. Note the coupling between the delay and scale axes; this twisting of the ambiguity function normally suggests that the original signal was a linear frequency modulated (LFM) signal [Alt, Van2] but, as can be seen in Figure 6.6, the original signal was a short tone. The ambiguity function is similar to a LFM signal because the delay axis is being warped by the scale parameter and with the tone-like mother wavelet, scale maps to frequency, so the frequency appears to be a function of delay or time. Since there is not really any coupling between the delay and scaling in this signal, the form of the WBCAF in equation (6.10) seems to be undesirable for direct spatial mappings. However, under certain circumstances that are discussed in Chapter 3, the WBCAF have desirable qualities; this section emphasizes that the form of the wideband processor cannot be arbitrarily chosen. The form of the wideband processor (WBCAF versus wavelet transform) dictates the characteristics of the estimates.

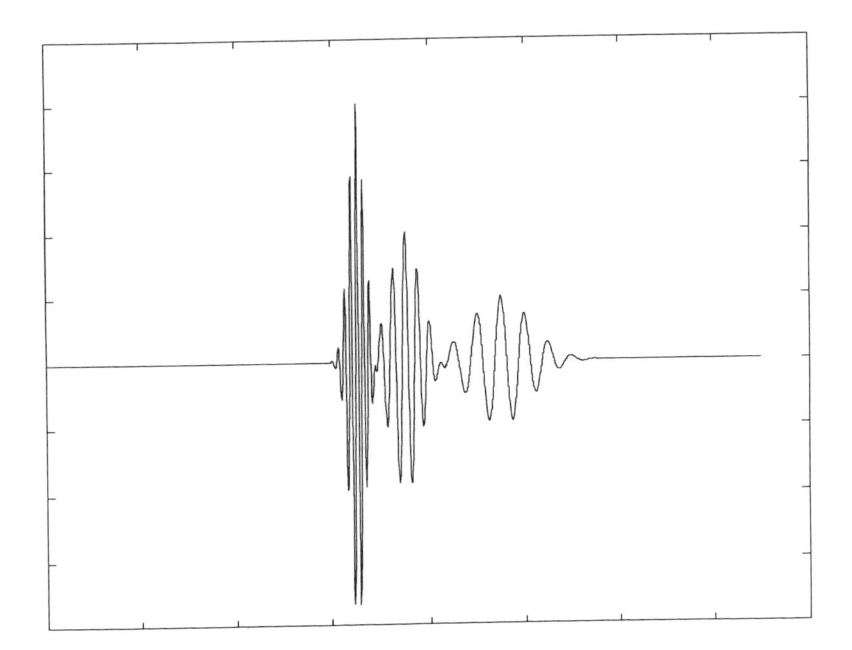

Figure 6.6: Received Signal

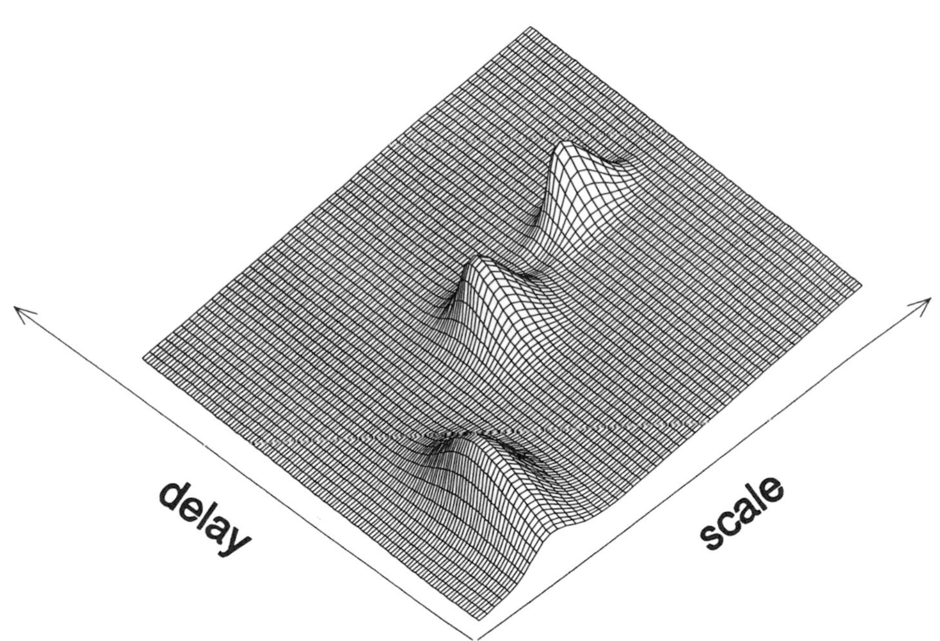

Figure 6.7: Magnitude of the Wavelet Transform of the Signal in Figure 6.6

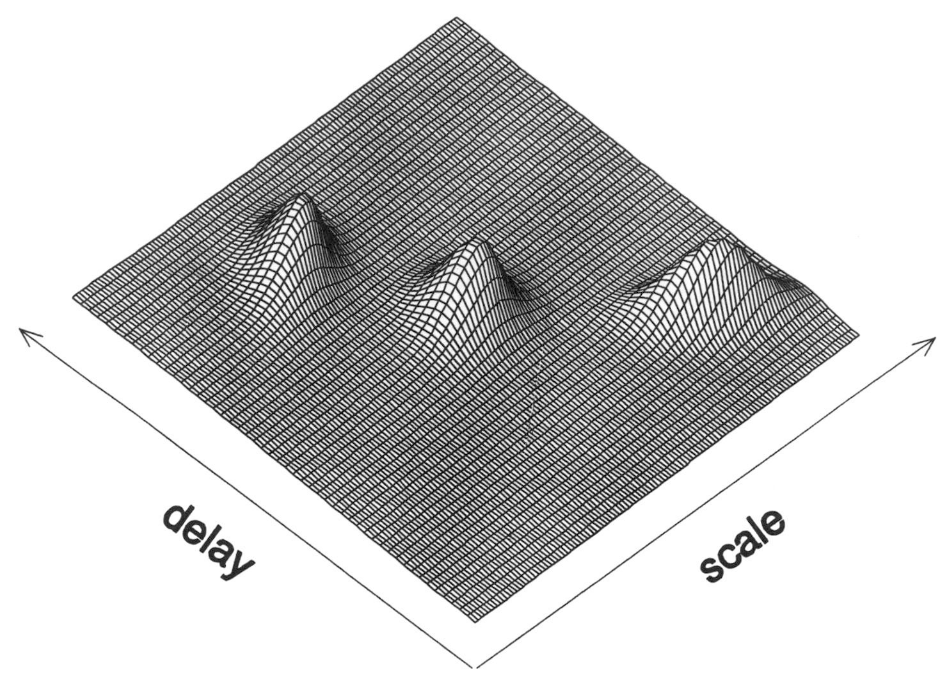

Figure 6.8: Magnitude of the WBCAF of the Signal in Figure 6.6

Time Delay and Scale Estimation

For the case of a single point scatterer in a simple environment (some assumptions were discussed earlier in this chapter), the peak of the wideband correlation surface can be mapped to a range and velocity. The variance of this positional estimate can be approximated from the variance of the delay-scale estimate. The form of the wavelet transform allows the variances in the scale-translation plane to be mapped *directly* to the range-velocity variances. The WBCAF in equation (6.10) would require its variances (actually the entire ambiguity function) to be warped in the translation-scale plane before they could be used for uncoupled range-velocity variance estimates.

Under the previously stated assumptions for the medium in the Point Scatter Example, the equations that relate the range to the delay and the radial velocity to scale are (also presented in Chapter 1, but repeated here for convenience):

$$range = \frac{c}{2} b \quad \text{and} \quad vel \approx \frac{c}{2}(a-1) = \frac{c}{2} a - \frac{c}{2} \qquad (6.12)$$

These equations are justified by considering the wavelet transform processor structure and the physical environment. The range to the scatterer changes over the duration of the signal for the wideband or wavelet case. Thus, the characterization of the environment must be referenced to some absolute time reference. As in the system modelling case, let the beginning of the received signal block be the chosen time reference. Let the delay be represented by the round trip delay time of this front edge of the signal (the signal is assumed to be in this block of data). Note that the velocity or scale will affect the delay parameter with the WBCAF definition because the delay is in "scaled time." Since the beginning of the block is defined as the time reference, then the time scaling acts relative to this time origin. When this entire block is scaled, the front edge of the transmitted signal moves due to the scaling (only in the WBCAF formulation). So, if the scaling acts first on the received signal block, then the delay is affected by the scaling. Therefore, the WBCAF in equation (6.10) couples the range and velocity estimates from the delay-scale estimates. To obtain uncoupled estimates that allows delay to map directly to the range of the scatterer the wavelet transform should be used.

The variance of the wavelet transform is assumed to map directly to *range-velocity uncertainties of the scatterers* *in the medium*. The reference frame chosen for analysis is critical. *The medium must be chosen as the reference frame.* The justification is that the *measurement uncertainty is in the medium or the range-velocity uncertainty.* Thus, the *time signals in the medium must be used as the reference signals*, not the signal sent to the transmitter. *The reference signal corresponds to the signal in the ambiguity function formulation whose argument is just time (corresponding to the time differential over which the integration occurs).*

Medium reference signals refer to those signals which were actually sensed in the medium. For the active processing, in which a signal is transmitted and reflected, the medium reference signal can only be the received signal. The time scale that is measured corresponds to the dilation or compression that happened to the transmitted signal. The transmitted signal should be scaled and delayed to look like the received signal.

The new interpretation is that the shape of wideband ambiguity function (and the positional uncertainty of the scatterer) changes as the scale of the reference signal changes. So the resolution or estimator performance of the wideband correlation receiver must change across scale. This is perfectly analogous to the resolution properties of wavelet transforms [Dau]. The wavelet transform resolution properties are the wideband *auto*-ambiguity function resolution properties.

WBAAF Moments and Assumptions

Another problem with wideband ambiguity function performance analysis is the invalid assumption that the form or shape of the ambiguity function does not change as it is translated across the ambiguity plane. The assumption is often termed:

"...without loss of generality, the performance analysis is performed around scale=1 and delay=0,"

This statement is invalid for wideband processing. The resolution properties of the wideband correlation receiver changes as the position of the wideband auto-ambiguity function of the transmitted signal changes as the scale is changed; it has constant shape only in a warped space, not the linear space of scale and delay as was demonstrated previously. Several researchers have used the aforementioned *invalid* invariance properties to design signals and filtering systems. This research should be readdressed with the warping property of wideband systems and signals being considered.

Wideband Scattering Review

The new system operator formulated in Chapter 5 is exploited to model wideband signals, systems, and scattering (spreading) with wavelet theory. Nonstationary systems or channels can be characterized even when the input and/or output signals are also nonstationary. The relationships to classical system characterizations, wavelet transforms and wideband cross ambiguity functions are detailed.

The estimation of the wideband scattering function is formed with a wideband correlation receiver. The characteristics of this correlator output are interpreted with ambiguity function analysis. The utilization of sophisticated signals does not limit these models as it does for the narrowband model. However, these models still only model linear time variation or constant radial velocities. Accelerations and higher order time mappings limit the estimation interval for these models, but this is an extension over the narrowband models which were limited by velocities. In addition, the bi-wavelet distribution of Chapter 5 does address a "double" linear time mapping that accounts for the acceleration.

As stated in Chapter 5, the new wideband/wavelet model attempts to estimate the environment (image) at one snapshot instant of time but processes the signals over a time interval (the estimation interval - often the length of the signal being processed). *It is assumed in the wideband assumption that the environment may have a different image or status at the end of the estimation interval than it did at the beginning of the estimation interval - the estimator accounts for the linear time variation over the estimation interval.* The narrowband models assume that the environment has not changed over the estimation interval. The narrowband assumption can become invalid over a shorter period of time than the wideband model. The extra estimation time achieved by employing the wideband model can lead to more robust estimates, greater energy, higher gains, better resolution and other desirable characteristics.

Related Research

This section is intended for the researcher who is already familiar with the wavelet theory or for the scientist/engineer who wants to see related references and how they relate to this book.

Disclaimer: The connections to related research considers only research that was directly exploited to write this book. These connections are only from the author's perspective and is biased by his own experiences and colleagues. Please accept an apology for works that have been overlooked or are uncited. Please forward suggestions and/or comments to the author.

Although much of the underlying mathematics and concepts of wavelet theory has been available for many years [Schur, Wigner, Ville, Duffin, Schaeffer, Gabor, Calderon, and many others] wavelet theory (that bears its name due to its initial application in seismology) started in France in the early 1980's in work by Grossman and Morlet [Gro, Mor]. This initial research utilized nonorthogonal wavelets exclusively. Several researchers including Battle and Meyer [Bat, Mey] quickly refined the mathematics and posed several problems related to this new concept of wavelet "bases." The field grew rapidly and became a world-wide research topic of many scientists, professors, and, thus, graduate students, including the author.

A critical question dominated all others: could an orthogonal wavelet basis be created? Battle, Lemarie, Meyer, Daubechies, Mallat [Bat, Mey, Dau, Mal] and many others attacked the problem. Affirmatively answering the question, researchers constructed orthogonal bases with several different types of constraints and construction algorithms and produced several different orthogonal wavelet bases. Some orthogonalization techniques introduced sensitivity, and regularity conditions had to be introduced. Daubechies [Dau] derived a constructive example of an orthogonal basis with finite duration basis functions. However, all of these initial orthogonal approaches operated on a dyadic grid (lattice of scale and translation points) - the wavelets are designed to be orthogonal on a pre-specified grid. Meyer and others also considered non-dyadic grids. The orthogonal conditions require specific mother wavelet designed for a specific grid.

In the late 1980's, mathematically complete discrete (in the wavelet domain) wavelet transform theory was constructed [Dau, Mal, Mey]. The discrete wavelet transform is often referred to as the wavelet series to be analogous to the Fourier series (continuous time functions but discrete frequencies). The mathematical theory of frames [Daub, Duff, Hei] and functional analysis were exploited to construct the constraints and the inverse discrete wavelet transform. The limitations on acceptable mother wavelets and wavelet domain grid densities were established [Coi, Dau, Mal, Mey]. An excellent tutorial paper by Heil and Walnut integrated group theory mathematics with the wavelet theory [Hei].

In the same time frame, multiresolution transforms were introduced by Mallat and Meyer [Mal, Mey]. Mallat related wavelet theory to the pyramidal decomposition of images by Burt, et. al. [Bur] which operated on discrete image data (pixels). For discrete data, multiresolution transforms were quickly related to multirate filtering banks [Vai]. The connections to subband coding, perfect-reconstruction quadrature mirror filters [Rio, Vet], and implementations with the finite duration orthogonal wavelets [Dau] followed rapidly. Multiresolution transforms were designed to operate on discrete time data - continuous time signals were not necessarily approximated. The efficiency improvements of multiresolution transforms directed their first application to be image processing. Image analysis, compressing (via subbanding), and quantization were, and continue to be, critical applications of the wavelet theory.

More specific to this book, after a non-technical introduction to wavelet theory in Chapter 1, Chapter 2 reviews established wavelet theory and is primarily a re-statement of other researchers work, reviewing the definitions, requirements, constraints, and limitations of the variety of wavelet transforms and reconstruction formulas. Most of this work is extracted from Grossman, Morlet, Daubechies, Mallat, and the edited book by Combes. The relationships between matched filters and wavelet transforms are detailed. Besides matched filtering, the book emphasizes the generality of wavelet transforms by changing the mother wavelet and the broad freedom in choice of a mother wavelet. The justification for stressing these topics is the later chapters of the books where very general mother wavelets are required.

Chapter 3 considers the resolution properties of wavelet transforms and the parameterization of the wavelet transforms by the mother wavelet. The mother wavelet controls the characteristics of a wavelet transform as was visually initially demonstrated by Grossman, Kronland-Martinet, and Morlet in [Com]. Resolution and, thus, efficiency of a signal representation depend upon the mother wavelet as well. These properties are most easily described and most broadly understood when they are related to matched filters and, thus, auto ambiguity functions (or uncertainty or point-spread functions) [Van]. Exploiting the related narrowband matched filter theory, the wideband or wavelet tradeoffs for wavelet domain grid densities can be examined more intuitively without being side-tracked by detailed mathematics.

In addition, my colleagues at Penn State, Sibul and Ricker have extensive experience with wideband ambiguity functions. Since wavelet transforms and wideband ambiguity functions are closely related, the two theories were rapidly fused. Prior research by Sibul and Titlebaum had even derived the admissibility condition as a requirement for one of the wideband signals being processed. For that application the admissibility condition was required so that the volume of the ambiguity function would not approach infinity [Sib]. Many related theories of wideband ambiguity functions apply to wavelet theory [Alt, Ric]. The mathematics of group theory and unitary representations integrate much of the two theories and has been brilliantly taught and addressed by Chaiyasena [Chai].

Sibul suggested utilizing the new wavelet transforms for wideband space-time processing. With the many practical problems associated with *true* wideband array processing (see Chapter 3), further investigations require a simultaneous, practical, and theoretical viewpoint. The extensions of the theory necessarily considered efficiency of implementation. Although orthogonal transforms offered efficiency, the standard multirate filtering implementation structures, and especially dyadic grids, do not provide the fine scale resolution that is required to represent systems with wavelet theory.

The requirements and constraints of the orthogonal case and the generality of the nonorthogonal case directed our research to the unconstrained wavelet case. Chapter 4 considers general signal processing applications where dyadic grids, real wavelets, and constrained mother wavelets; these properties may be undesirable in some applications and cannot be tolerated in others. For some problems, freedom in choice of the mother wavelet and its resolution properties is important (one particular application of representing systems is detailed in Chapter 5). In other problems, multiple mother wavelets can more efficiently represent the signal than a single mother wavelet. An added dimensionality of the wavelet transform over the Fourier transform is the freedom in choosing the kernel of the integral transform (the kernel can be almost any bandpass signal). To exploit this freedom and maintain efficient computation, wavelet theory was extended by the new "mother mapper operator" [You]. A precursor derivation, which is more physically understood and related to established structures, constructs the wideband cross ambiguity function in *space* with the wavelet transform of two *time domain* signals. The mapping of two time domain signals to spatial coordinates has been performed in the narrowband case by Chestnut [Che] and motivates the space-time processing performed with wavelet theory. Several general applications of the mother mapper operator are presented. This research is most closely related time-varying systems and narrowband scattering theory, but since Chapters 5 and 6 cover these exclusively, the specific references are cited there.

Besides the array processing, Chapter 5 addresses general system theory. Sibul, Bose, and Ricker suggested applying the new wavelet operator model space-time-varying systems. The result was the STV (space-time-varying) operator that maps and a space or time domain input to a space or time domain output. However, with the STV operator, the system can be space and/or time varying and the inputs and outputs can be wideband and nonstationary. More significantly, the operator is simple and well characterized; it is essentially an inverse wavelet transform *with the input acting as the mother wavelet* and the reconstructed signal being the system output. Note the requirement for very general mother wavelets. Several related models have been proposed that used theories related to wavelet theory. Narparst and O'Sullivan [Nar, O'S] have specifically addressed the scattering problem with wavelet theory (although they used a different set of assumptions than used in this book and did not address the general system problem - just scattering). Drumheller [Dru1] considered the action of a LTI system on signals represented with wavelet transforms. Several others considered representing systems with Wigner theory [Hla].

Finally, Chapter 6 considers the application of the STV wavelet operator system model for the "scattering problem" and the related wideband signal design problem. Narrowband scattering theory and its relationship to ambiguity functions has been extensively addressed [Spa, Van, Woo, Zio]. The research of Altes, Ricker, Titlebaum, Ziomek, and many others addresses the extension of the wideband case from the narrowband concepts. Since wideband scattering theory is more general, the theory should start at the wideband case and be further constrained to arrive at the narrowband case, but due to many considerations that has not occurred (in addition, the most general case includes higher order time-mapping terms such as acceleration). However, Chaiyasena and Sibul [Cha] have started from the wideband and "contracting" the mapping to the narrowband case.

Since the system model is necessary to determine the scattering function in the narrowband case [Van, Zio], why not first address the wideband system model; this is the approach taken. Due to several conditions, the wideband scattering is not modelled by a magnitude-only function as in the narrowband case. In addition, unlike the invariance of the narrowband ambiguity function across the time-frequency plane the wideband ambiguity function is not invariant across the time-scale plane. For these reasons and others, the coherent STV wavelet system model, the wideband system characterization of Chapter 5, is stated as the wideband scattering function. Justification for a specific form of the wideband ambiguity function are also stated [Alt, Kel, Ric, Zio]. Finally, wideband signal design is briefly addressed to resolve some previous research [Alt, Ric, Tit] and attempt to direct future research.

Many of the new concepts and interpretations presented in this book were inspired by Prof. L. H. Sibul. In addition he helped polish the presentation of the book. Prof. N. K. Bose also helped develop these ideas as my Ph.D. thesis advisor. Finally, the author wishes to thank his friends and colleagues for their contributions and significant refinements to this book: L. Weiss, K. Hillsley, A. P. Chaiyasena, T. Dixon, M. Fowler, and many unmentioned others.

References

Note that some authors are referenced multiple times in this list. If a concept in the book can be referenced in more than one of the author's papers, then the reference is simply the first three letters of the author's name rather than listing all of the specific references.

[Alt1] Altes, R. A., "Method of wideband signal design for radar and sonar systems," Ph.D. Dissertation, Dept. Elect. Engr., Univ. of Rochester, Rochester, NY, 1970

[Alt2] Altes, R. A., "Some invariance properties of the wide-band ambiguity function," *JASA*, vol. 53, no. 4, 1973, pp. 1154-60

[Alt3] Altes, R. A., "Target position estimation in radar and sonar, generalized ambiguity analysis for maximum likelihood parameter estimation," *Proc. IEEE*, vol. 67, 1979, pp. 920-930

[Alt4] Altes, R., "Wavelets, tomography, and line segment image representations," *Proc. SPIE* 1348, 1990, pp. 268-278

[Aus1] Auslander, L., and Gertner, I., "Wide-band ambiguity function generation and ax+b group," from *Signal Processing, Part 1: Signal Processing, vol. 1*, Springer-Verlag, New York, 1990, pp. 1-12

[Aus2] Auslander, L., and Tolimiere, R., "Radar ambiguity functions and group theory," *SIAM J. Math. Anal.*, vol. 16, no. 3, May 1985, pp. 577-601

[Bas] Basseville, M., and Benveniste, A., "Multiscale statistical signal processing," *IEEE ICASSP*, 1989, pp. 2065-2068

[Bat] Battle, G., "A Block Spin Construction of Ondelettes (wavelets), Part 1: Lemarie functions," *Comm. Math. Physics*, vol. 110, 1987, pp. 601-615

[Bel] Bello, P. A., "Characterization of Randomly Time-variant Linear Channels," *IEEE Trans. Communication Sys.*, vol. 11, 1963, pp. 360-393

[Bos1] Bose, N. K., *Digital Filters Theory and Applications*, Elsevier Science Publishing, New York, 1985

[Bos2] Bose, N. K. (ed.), *Multidimensional Systems Theory: Progress, Directions and Open Problems*, D. Reidel Publishing Co., Dordrecht, Holland, 1985

[Bou] Boudreaux-Bartels, G. F., "Time-Varying Signal Processing Using the Wigner Distribution Time-Frequency Signal Representation," *Advances in Geophysical Data Processing*, vol. 2, Jai Press Inc., 1985, pp. 33-79

[Bur] Burt, P. J., and Adelson, E. H., "The Laplacian Pyramid as a Compact Image Code," *IEEE Trans. Communications*, vol. 31, no. 4, 1983, pp. 532-540

[Cap] Capon, J., "High Resolution Frequency Wavenumber Spectrum Analysis," *Proc. IEEE*, vol. 57, no. 8, August 1969, pp. 1408-1418

[Car1] Carter, G. C., and Abraham, P. B., "Estimation of Source Motion from Time Delay and Time Compression Measurement," *JASA* vol. 67, no. 3, March 1980, pp. 830-832

[Car2] Carter, G. C. (editor), "IEEE Special Issue on Time Delay Estimation," *IEEE Trans. Acoust. Speech Signal Proc.*, vol. 29, no.3, June 1981

[Chai] Chaiyasena, A. P., Sibul, L. H., and Banyaga, A., "The Relationship Between Narrowband and Wideband Ambiguity Volume Properties: A Group Contraction Approach," Conference on Information Science and Systems, Johns Hopkins, Baltimore, March 1991

[Cham] Champagne, B., Eizenman, M. and Pasupathy, S., "Factorization Properties of Optimum Space-Time Processors," *IEEE Trans. Acoust. Speech Signal Proc.*, vol. 38, no. 11, Nov. 1990, pp. 1853-1869

[Chan] Chan, Y. T., Riley, J. M., and Plant, J.B., "Modeling of Time Delay and Its Application to Estimation of Nonstationary Delays," *IEEE Trans. Acoust. Speech Signal Proc.*, vol. 29, no. 3, June 1981, pp. 577-581

[Che] Chestnut, P. C., "Emitter Localization Accuracy Using TDOA and DifferentialDoppler," *IEEE Trans. Aero. Elect. Systems*, vol. 18, no. 2, March 1982, pp. 214-218

[Chu] Chui, C. K., *An Introduction to Wavelets*, Academic Press, Inc., San Diego, 1992

[Cla] Classen, T., and Mecklenbrauker, W., "The Wigner distribution--a tool for time-frequency signal analysis," Phillips J. Res. 35, 1980, pp. 217-250, pp. 276-300, pp. 372-389

[Coh] Cohen, L., "Time-frequency distribution--A review," *Proc. IEEE* 77-7, 1989, pp. 941-981

[Coi] Coifman, R. R., "Wavelet Analysis and Signal Processing," in *Signal Processing, Part I: Signal Processing Theory*, L. Auslander et al. eds., IMA, vol. 22, Springer, New York, 1990

[Com] Combes, J. M., Grossman, A., and Tchamitchian, Ph., *Wavelets: Time-Frequency Methods and Phase Space*, Springer-Verlag, New York, NY, 1989

[Cro] Crochiere, R. E., Rabiner, L. R., *Multirate Digital Signal Processing*," Prentice-Hall, Inc., Englewood Cliffs, NJ, 1983

[Dau1] Daubechies, I., "Orthonormal bases of compactly supported wavelets," *Comm. Pure Appl. Math.*, vol. 41, 1988, pp. 909-996

[Dau2] Daubechies, I., "Time-frequency localization operators: a geometric phase space approach," *IEEE Trans. Inform. Theory*, vol. 34, 1988, pp. 605-612

[Dau3] Daubechies, I., "The wavelet transform, time/frequency localization and signal analysis," *IEEE Trans. Inform. Theory*, vol. 36, Sept. 1990, pp. 961-1005

[Dau4] Daubechies, I., "The Wavelet Transform: A Method for Time-frequency Localization," Chapter 8 of *Advances in Spectrum Analysis and Array*

Processing: vol. 1, (edited by Simon Haykin) Prentice Hall, Englewood Cliffs, NJ, 1991, pp. 366-417

[Dru1] Drumheller, D. M., "Theory and Application of the Wavelet Transform to Signal Processing," NRL Report 9316, July 31, 1991

[Dru2] Drumheller, D. M., "A Computer Program for the Calculation of Daubechies Wavelets," NRL Report 6923, Jan. 22, 1992

[Duf] Duffin, R. J., and Schaeffer, A. C., "A Class of Nonharmonic Fourier Series," *Trans. Am. Math. Soc.*, vol. 72, 1953, pp. 341-366

[Fla1] Flandrin, P., "Wavelets and related time-frequency transforms," *Proc. SPIE* 1348, 1990, pp. 2-13

[Fla2] Flandrin, P., and Rioul, O., "Affine smoothing of the Wigner-Ville distribution," *IEEE Int. Conf. on Accoust., Speech, and Signal Proc.*, 1990, pp. 2455-2486

[Fol] Folland, G. B., *Harmonic Analysis on Phase Space*, *Annals of Math. Studies*, no. 122, Princeton University Press, Princeton, N.J., 1989

[Fow] Fowler, M. L., and Sibul, L. H., "Signal Detection Using Group Transforms," *IEEE ICASSP*, Toronto, May 1991, pp. 1693-1696

[Gab] Gabor, D., "Theory of Communication," *IEE Proc.* (London) vol. 93, no. 3, Nov. 1946, pp. 429-457

[Gli] Glisson, T. H., and Sage, A. P., "On Sonar Signal Analysis," *IEEE Trans. Aero. Elect. Systems*, vol. 6, no. 2, Jan. 1970, pp. 37-49

[Gro1] Grossmann, A., and Morlet, J., "Decomposition of Hardy functions into square integrable wavelets of constant shape," *SIAM J. Math. Anal.*, vol. 15, 1984, pp. 723-736

[Gro2] Grossmann, A., Morlet, J., and Paul, T., "Transforms associated to square integrable group representations, I, General results," *J. Math. Physics*, vol. 26, 1985, pp. 2473-2479

[Gro3] Grossmann, A., Morlet, J., and Paul, T., "Transforms Associated to Square Integrable Group Presentations II: Examples," *Am. Inst. Henri Poincare*, vol. 45, no. 3, 1986, pp. 293-309

[Hah] Hahn, W. R., "Optimum Signal Processing for Passive Range and Bearing Estimation," *JASA* vol. 58, no. 1, July 1975, pp. 201-207

[Has] Hassab, J. C., and Boucher, R. E., "Optimum Estimation of Time-Delay by a Generalized Correlator," *IEEE Trans. Acoust. Speech Signal Proc.*, vol. 27, no. 4, Aug. 1979, pp. 373-380

[Hei] Heil, C. E., and Walnut, D. F., "Continuous and Discrete Wavelet Transforms," *SIAM Review*, vol. 31, no. 4, Dec. 1989, pp. 628-666

[Hla] Hlawatsch, F., and Boudreaux-Bartels, G. F., "Linear and Quadratic Time-frequency Signal Representations," *Signal Processing Magazine*, vol. 9, no. 2, April. 1992, pp. 21-67

[IEEE] Special Issue on Wavelet Transforms and Multiresolution Signal Analysis, *IEEE Trans. Inform. Theory*, vol. 38, no. 2, 1992

[JohB] Johnson, B. L., "Scattering Function Identification Via Constrainted Iterative Deconvolution," Ph. D. Dissertation, Acoustics Dept., The Penn. State Univ., Univ. Park, PA, 1992

[JohD] Johnson, D. H., "The Application of Spectral Estimation Methods to Bearing Estimation," *Proc. IEEE*, vol. 70, 1982, pp 1018-1028

[Kel] Kelly, E. J., and Wishner, R. P., "Matched-Filter Theory for High-velocity Targets," *IEEE Trans. Military Elect.*, vol. 9, 1965, pp. 56-69

[Kla] Klauder, J. R., and Sudarshan, E. C. G., *Fundamentals of Quantum Optics*, W. A. Benjamin, Inc., New York, 1968

[Kna1] Knapp, C. H., and Carter, G. C., "The Generalized Correlation Method for Estimation of Time Delay," *IEEE Trans. Acoust. Speech Signal Proc.*, vol. 24, Aug. 1976, pp. 320-327

[Kna2] Knapp, C., and Carter, G. C., "Estimation of Time Delay in the Presence of Source or Receiver Motion," *JASA*, vol. 66, no. 6, June 1977, pp. 1545-1549

[Kni] Knight, W. C., Pridham, R. G., and Kay, S. M. "Digital Signal Processing for Sonar," *Proc. IEEE*, vol. 69, no. 11, 1981, pp. 1451-1506

[Kuo] Kuo, B. C., *Automatic Control Systems*, Prentic-Hall, Inc., Englewood Cliffs, NJ, 1987

[Mal1] Mallat, S. G., "A theory for multiresolution signal decomposition: the wavelet representation," *IEEE Trans. Pattern Anal. Machine Intel.*, vol. 31, 1989, pp. 674-693

[Mal2] Mallat, S. G., "Multiresolution approximation and wavelets," *Trans. Am. Math. Soc.*, vol. 135, 1989, pp. 69-88

[Mal3] Mallat, S. G., "Multifrequency channel decompositions of images and wavelet models," *IEEE Trans. Acoust. Speech Signal Proc.*, vol. 37, 1989, pp. 2091-2110

[Mey] Meyer, Y., *Ondelettes*, Hermann, Paris, 1990

[Nap] Naparst, H., "Dense Target Signal Processing," *IEEE Trans. Inform. Theory* vol. 37, no. 2, 1991, pp. 317-327

[O'S] O'Sullivan, J. A., "A Transform Approach to Radar Imaging," submitted to the *IEEE Trans. Inform. Theory*, July, 1991

[Ows] Owsley, N., and Swope, G., "Time Delay Estimation in a Sensor Array," *IEEE Trans. Acoust. Speech Signal Proc.*, vol. 29, no. 3, June 1981, pp. 519-523

[Pap1] Papoulis, A., *Probability, Random Variables, and Stochastic Processes*, McGraw-Hill Book Co., New York, 1984

[Pap2] Papoulis, A., *Signal Analysis*, McGraw Hill Book Co., New York, 1977

[Pre] Preistley, M. B., *Non-linear and Non-stationary Time Series Analysis*, Academic Press, San Diego, 1989

[Ric1] Ricker, D. W., "Small Aperture Angle Measurement for Active Echo Location Systems," *IEEE Trans. Aero. Elect. Systems*, vol. 22, no. 4, July 1986, pp. 380-388

[Ric2] Ricker, D. W., "The Doppler Sensitivity of Large TW Phase Modulated Waveforms," *IEEE Trans. Signal Proc.*, Oct. 1992

[Rio] Rioul, O., and Vetterli, M., "Wavelets and Signal Processing," *Signal Processing Magazine*, vol. 8, no. 4, October, 1991, pp. 14-38

[Sch1] Schempp, W., "Radar ambiguity functions, the Heisenberg group, and holomorphic theta series," *Proc. Am. Math. Soc.*, vol. 92, 1984, pp. 103-110

[Sch2] Schempp, W., "Analog radar design and digital signal processing - a Heisenberg nilpotent Lie Group approach," Chapter 1 of *Lie Methods in Optics*, (edited by J. Sanchez-Mandragon and K. B. Wolf) Springer, New York, 1986, pp. 1-27

[She] Shensa, M. J., "The Discrete Wavelet Transform: Wedding the a Trous and Mallat Algorithms," submitted to *IEEE Trans. Acoust. Speech Signal Proc.*, 1990

[Sib1] Sibul, L. H., Chaiyasena, A. P., and Fowler, M. L., "Signal Ambiguity Functions, Wigner Transforms, and Wavelets," in *Signal Processing and Digital Filters*, M. H. Hamza, editor, Lugano, Switzerland, June 1990, pp. 214-217

[Sib2] Sibul, L. H., and Titlebaum, E. L., "Volume Properties for the Wideband Ambiguity Function," *IEEE Trans. Aero. Elect. Systems*, vol. 17, 1981, pp. 83-86

[Sib3] Sibul, L. H., and Ziomek, L. J., "Generalized Wideband Crossambiguity Function," *IEEE ICASSP*, 1981, pp. 1239-1242

[Smi] Smith, M. J. T., and Barnwell, T. P., "Exact Reconstruction for Tree-Structured Subband Coders," *IEEE Trans. Acoust. Speech and Signal Proc.*, vol. ASSP-34, pp. 434-441

[Spa] Spafford, L. J., "Optimum Radar Signal Processing in Clutter," *IEEE Trans. Inform. Theory*, no. 14, 1968, pp. 734-743

[Spe] Speiser, J. M., "Wide-Band Ambiguity Functions," *IEEE Trans. Inform. Theory*, pp. 122-123, 1967

[SteS] Stein, S., "Algorithms for ambiguity function processing," *IEEE Trans. Acoust. Speech Signal Proc.*, vol. 29, 1981, pp. 588-599

[SteB] Steinberg, B. D., *Principles of Aperture and Array System Design*, Wiley, 1976

[Str] Strang, G., "Wavelets and dilation equations: A brief introduction," *Siam Rev.* 31-4, 1989, pp. 614-627

[Swi] Swick, D. A., "A Review of Wideband Ambiguity Functions," NRL Report 6994, 1969

[Vai1] Vaidyanathan, P. P., "Multirate Digital Filters, Filter Banks, Polyphase Networks, and Applications: A Tutorial," *Proc. IEEE*, vol. 78, no. 1, 1990, pp. 56-93

[Vai2] Vaidyanathan, P. P., "Quadrature Mirror Filter Banks, M-band Extensions and Perfect-Reconstruction Techniques," *IEEE ASSP Magazine*, vol. 4, no. 3, pp. 4-20, July 1987

[Vak] Vakman, D. E., *Sophisticated Signals and the Uncertainty Principle in Radar*, Translated from Russian by K. N. Trirogoff, New York: Springer-Verlag, 1968

[Van1] Van Trees, H., *Detection, Estimation and Modulation Theory, Part I*, Wiley, 1968

[Van2] Van Trees, H., *Detection, Estimation and Modulation Theory, Part III*, Wiley, 1971

[Vet1] Vetterli, M., and Herley, c., "Wavelets and Filter Banks: Theory and Design," Report CU/CTR/TR 206/90/36, Center for Telecommunications Research, Columbia Univ., NY, NY, 1990

[Vet2] Vetterli, M., "A theory of multirate filter banks," *IEEE Trans. Acoust. Speech, Signal Proc.*, vol. 35, 1987, pp. 356-372

[Vet3] Vetterli, M., "Filter Banks Allowing Perfect Reconstruction," *Signal Processing*, vol. 10, no. 3, April 1986, pp. 219-244

[Woo] Woodward, P. M., *Probability and Information Theory with Applications to Radar*, Oxford, England:Pergamon, 1964

[Yag] Yaglom, A. M., *Theory of Stationary Random Functions*, Prentice-Hall, 1962

[YouR] Young, R., *An Introduction to Nonharmonic Fourier Series*, Academic Press, New York, 1980

[You1] Young, R. K., Sibul, L. H., and Bose, N. K., "Wideband Space-Time Processing and Wavelets," presentation and paper in the Conference on Information Science and Systems, Johns Hopkins, Baltimore, March 1991

[You2] Young, R. K., "Wideband Space-Time Processing and Wavelet Theory," Ph. D. Dissertation, Dept. Elect. Engr., The Penn. State Univ., Univ. Park, PA, 1991

[You3] Young, R. K., "Wavelet Transform Mother Mapper Operator", Asilomar Conf. on Signals, Systems, and Computers, Nov. 1991, Pacific Grove, CA

[Zad] Zadeh, L. A., and Desoer, C. A., *Linear System Theory: Thes State Space Approach*, McGraw-Hill, New York, 1963

[Zho] Zhong, S., "Complete Signal Representation with Multiscale Edges," submitted to *IEEE Trans. Pattern Anal. Machine Intel.*, 1989

[Zio] Ziomek, L. J., *Underwater Acoustics: A Linear System Theory Approach*, Academic Press, New York, 1985

Subject Index